*Celestial Sites, Celestial Splendors* presents us with a beautiful and fascinating panorama of more than a hundred carefully chosen astronomical objects – from nearby planets, meteors and comets, to distant stars, galaxies and nebulae – and provides a comprehensive and practical guide to viewing them. Each object's location , discovery and optimal observation times are discussed, along with advice on the best techniques for observing it. We are shown how these splendors of the sky can be seen with simple instruments, such as binoculars or a small telescope, and sometimes even with the naked eye. The book also gives advice on the best choice of accessories to use with amateur telescopes and contains photographs illustrating how each object appears when seen through such apparatus, along with excellent pictures from large space telescopes. Its accessible style will appeal to all amateur astronomers and anyone interested in the night sky.

Prior to opening his own bookshop HERVÉ BURILLIER ran, for seven years, the biggest astronomy specialist bookshop in France, stocking over five hundred titles. He is a member of the French Astronomical Society (SAF) and of the French Association of Variable Star Observers, and has written the column 'Observing the sky' in *L'Astronomie* (the monthly publication of the French Astronomical Society) since 1990. He was honoured by the Foundation of France in 1991 and decorated by the French Astronomical Society in 1994.

# CELESTIAL
## SITES, CELESTIAL
# SPLENDORS

## HERVÉ BURILLIER

Translated by Nathalie Audard-Sword

### CAMBRIDGE
### UNIVERSITY PRESS

PUBLISHED BY THE PRESS SYNDICATE OF THE UNIVERSITY OF CAMBRIDGE
The Pitt Building, Trumpington Street, Cambridge, United Kingdom

CAMBRIDGE UNIVERSITY PRESS
The Edinburgh Building, Cambridge CB2 2RU, UK   http://www.cup.cam.ac.uk
40 West 20th Street, New York, NY 10011–4211, USA   http://www.cup.org
10 Stamford Road, Oakleigh, Melbourne 3166, Australia
Ruiz de Alarcón 13, 28014 Madrid, Spain

Previously published in French as Les Plus Belles Curiosités Célestes by
Hervé Burillier © Bordas, Paris 1995 © Larousse Bordas, Paris 1998.
English edition © Cambridge University Press 2000

First published 2000

Printed in the United Kingdom at the University Press, Cambridge

Typeface Meta 9/12 pt    System QuarkXPress® [DS]

*A catalogue record for this book is available from the British Library*

*Library of Congress Cataloguing in Publication data*
Burillier, Herv, 1967–
Celestial sites, celestial splendors/Hervé Burillier; translated by Nathalie
Audard-Sword.
p. cm.
ISBN 0 521 66773 9 (pb)
1. Astronomy — Observers' manuals. I Title.
QB64.B82 2000
523 21–dc21    99-044952

ISBN 0 521 66773 9 paperback

# Contents

NB: The terms in bold on pages 2 to 10 correspond to the basic concepts defined in the introduction. In the rest of the book, the terms in bold are defined in the glossary (page 184).

Throughout this book, apertures of telescopes are quoted in *centimeters* or *millimeters*. To convert centimeters to *inches*, divide by 2.54. There are 10 millimeters in 1 centimeter. The symbol ∅ denotes the diameter.

# PREFACE

On a clear, dark, moonless night, the sky in all its splendor is transformed into a vast dome, studded with a myriad of sparkling points of light. Sometimes, a shooting star, a luminous arrow, appears from nowhere, interrupting the serenity of this eternal and unchanging scene, before dying and becoming only a memory.

Turn your gaze to the south, and watch the constellations every few hours, and see how they fall over the edge to the west, carrying with them the legends and myths of so many peoples.

Brave the January cold. The constellation Auriga reaches the celestial zenith of the winter sky, while to the south is the jewel of these freezing nights, Orion The Hunter, accompanied by his faithful dogs, Canis Major and Canis Minor. The giant is tireless in his pursuit of his neighbors Taurus and The Pleiades. Only the Scorpion frightens him and makes him run away.

In the May twilight, ruddy Antares begins to appear in the southwest. Then, Orion, sovereign of our cold winter skies, finally disappears below the western horizon with the setting Sun.

Meanwhile, climbing higher and higher, The Great Bear ascends towards the zenith. Capella shines golden. The Hyades and Aldebaran, flaming purple, plunge to the west and die with the Sun. Vega, a white and luminous star, appears in the northeast. Regulus crosses the meridian, preceded by the Crab, which was banished to the sky by the goddess Juno for having tried to bite the foot of Hercules while the hero was fighting the many-headed Hydra.

These are the ways of the sky, forever mute under our fascinated gaze ...

# ACKNOWLEDGEMENTS

The author acknowledges those people whose help, advice and support have brought this project to completion: G.-L. Carrié, J.-M. Lecleire, C. Lehénaff, J.-M. Lopez, J.-C. Merlin, Pises Observatory and the French Astronomical Society for the documents they provided; J. Minois for reading the manuscript and for his valuable comments; l'AFOEV and especially E. Schweitzer. The author and the editor thank the *Ciel et Espace* team, and Stéphane Aubin in particular, for their kind and efficient help, and Nathalie Audard-Sword for the English translation.

## Captions for the doubled-page chapter openings

**Pages 12–13**: *This artist's impression shows the Sun and its retinue of planets as a distant observer would see them* (© S. Numazawa/APB/*Ciel et Espace*).

**Pages 32–33**: *The Pleiades (or M45), the most famous open cluster, is typical of a young group of stars born from the same nebula, some 50 million years ago. These hot stars, beautifully bluish in color, still lie in the vast nebulosity in which they were born.* (© ROE/AAO/D. Malin/*Ciel et Espace*.)

**Pages 80–81**: *Unfortunately not observable from the Northern hemisphere, the globular cluster 47 Tucana is an exceptional concentration of stars next to the Small Magellanic Cloud* (© ESO/*Ciel et Espace*).

**Pages 110–111**: *The Helix Nebula, in the constellation Aquarius, is a planetary nebula. A hot star lies at the center of an expanding gas envelope, the last stage in the lives of giant stars and red supergiants. In a reflector, this type of nebulosity shows as a small disk, like a planet (which led to the misleading name, planetary nebulae).* (© ESO/*Ciel et Espace*.)

**Pages 130–131**: *The Orion Nebula (or M42) is an example of a diffuse nebula – a cloud of gas and interstellar dust illuminated by one or several stars located in or near its heart* (© AAO/D. Malin/*Ciel et Espace*).

**Pages 146–147**: *A region of sky near the center of the Virgo cluster* (© ROE/AAO/D. Malin/*Ciel et Espace*).

**Pages 168–169**: *Halley's Comet in the Milky Way* (© A. Fujii/*Ciel et Espace*).

# BASICS

The vault of the sky appears to rotate, reflecting the **diurnal motion** of the Earth (Figs. 1, 2, and 3). It takes 24 **sidereal hours** for a complete rotation (exactly 23 h 56 m 4 s). Celestial bodies therefore describe 360 degrees in 24 hours, that is, an arc equal to $\frac{1}{24}$ of a circle (or 15 degrees) in one hour.

This confirms the Earth's rotation, and our impression of the slow east to west motion of celestial bodies.

Imagine a line starting at the south pole, going through the center of the Earth, and coming out at the north pole. This line represents the **Earth's axis of rotation**.

Now extend this imaginary line towards the stars. It would go far beyond our planetary system, and eventually pass close to another sun. To the joy of astronomers, this

*Figure 1 illustrates the case of an observer located at the equator. In this case, the north and south celestial poles are on the horizon (H). All stars are visible. They rise in the eastern half of the sky, reach their highest point, then move westwards and set (in the direction of the colored arrows).*

*For an observer in Boston, at a latitude roughly half way up the northern hemisphere (fig. 2), the north celestial pole is above the horizon (H). The star E rises, reaches its highest point and sets, whilst star E', which is circumpolar, will be visible all night.*

*Now consider an observer at the north pole. The north pole is in the zenith. All the stars are circumpolar. They rotate, neither rising nor setting, always staying at the same altitude above the horizon. The cardinal points are no longer defined.*

other sun is a small, moderately luminous, star, located in the constellation Ursa Minor (also called the Little Bear).

This star was called the North Star (or Pole Star) by the first observers. Its location in the sky indicates the **north celestial pole**.

Also called Polaris, this star is not located exactly at the north pole. The distance which separates it nowadays from the actual north celestial pole is, nevertheless, smaller than one degree (46′ to be precise). The amateur astronomer need not be concerned about this difference, since the motion of the North Star around the true celestial pole is too small to be noticed during an observation. However, this motion must be allowed for in astronomical photography.

## Some terms to remember

In their motion around the north celestial pole, some stars describe complete circles. They stay above the horizon without ever rising or setting. These stars are said to be **circumpolar** and are always visible throughout the entire year (Fig. 4).

Other celestial bodies, stars, planets, and also the Sun and the Moon,

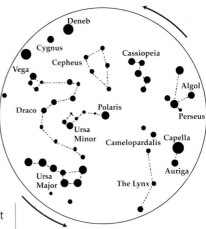

*Circumpolar stars and constellations visible from Europe.*

rise in the east and climb towards their highest point above the horizon. They are then said to cross the **meridian**. Next, they descend towards the western horizon and set. But, unlike the stars, which the ancients imagined to be attached to a **fixed sphere**, the planets, the Sun and the Moon rise and set at times which change with the seasons.

In winter, we can all see that the Sun rises late in the southeast, and sets early in the southwest. Each day after the winter solstice brings us closer to spring, and the points where the Sun rises and sets are shifted northward. Day after day the Sun climbs higher and higher in the sky, the days become longer and the nights shorter. At the summer sol-

3

stice this movement reaches its limit and then goes into reverse: the days become shorter, the nights longer. The Sun rises and sets a little more to the south each day until it is winter again. The years pass, with clockwork regularity.

The Moon follows the same laws. Our satellite moves rapidly in the sky; its position shifts by over 10 degrees a day and it therefore moves through all the constellations of the zodiac in just over 27 days.

Apart from its own rotation, the Earth also rotates around the Sun. This is called the **revolution of the Earth**: it takes 365.25 days to make a full orbit, at an average orbital speed of 30 kilometers or so per second. People have been fascinated by the measure of time since the beginning of history; we have drawn up our calendars according to these two motions: the rotation of the Earth, which regulates the length of day and night, and the revolution of the Earth around the Sun, which heralds the seasons and the passing of the years.

## Precessions and the ecliptic

In reality things are not as simple as they seem. Firstly, our planet is not a perfect sphere but bulges at its equator. This equatorial zone is subject to tidal forces from the Sun and the Moon, and the axis of rotation of the Earth describes a cone similar to that of a spinning top. This motion, called **precession**, is the cause of the shift of seasons over a very long time, and the combined effect of the Sun and the Moon is called **solar–lunar precession**. Because the Sun is very massive and the Moon is relatively close to the Earth, their combined forces provoke the rotation of the Earth's axis around an axis perpendicular to a plane called the **plane of the ecliptic**. This plane is defined by the apparent annual trajectory of the Sun on the celestial vault. It takes 26 000 years to complete the precession. Today, the terrestrial equatorial plane is at an angle of $23° 27'$ to the plane of the ecliptic. This angle, called the **obliquity of the ecliptic**, varies and is also subject to the perturbation forces of the Sun and the Moon. Its period is much longer – 40 000 years or so – and it varies between $21° 55'$ and $24° 18'$.

The Earth and the Moon are subject to many oscillations. We shall not describe them here, but simply note that they are due to continual differences in the direction of gravitational forces, i.e. differences in the respective positions of the Sun and of the Moon. These small oscillations are called **nutations**. The most important of these has a period of 18.6 years, and is the revolution of the lunar nodes (points where the orbit of the Moon crosses the ecliptic).

The combined effects of precession and nutation lead to complex motions of the celestial poles. Celestial mechanics is difficult, and we can appreciate the mental gymnastics made by the philosophers and scientists of the past in their attempts to understand it.

### Equatorial coordinates

In order to determine the position of a point on the surface of the Earth, of a town for example, we need to know two coordinates:

- the latitude of the town, i.e. its angular distance from the equator. We have adopted the following convention: we use negative latitudes in the southern hemisphere (−90° to 0°) and positive latitudes for regions in the northern hemisphere (0° to 90°);
- the longitude of the town. Longitude is the angular measure from

this point to a point of origin, agreed in 1887 to be the Greenwich Observatory, in England. We measure longitude from the reference meridian of Greenwich, positively westward, and negatively eastward.

To determine the position of an object in the sky, astronomers use a coordinate system similar to that used on Earth. Consider the celestial sphere rotating around the Earth's axis. By convention, we call the extension into space of the Earth's equatorial plane, at right angles to the Earth's axis, the **celestial equator**. This plane cuts the celestial sphere into two hemispheres: the **austral** hemisphere, from the south celestial pole to the celestial equator, and the **boreal** hemisphere, from the celestial equator to the north celestial pole.

We can now determine the position of any object in the sky. Consider the case of the star E (Fig. 5). The first measurement is the **declination**, also called the delta coordinate, or simply the Greek letter δ. Declination measures the angular distance between the object and the celestial equator and is given in degrees (°), arc minutes (′)

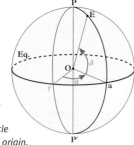

*Equatorial coordinates. Let PγP' be the hour circle defining the origin, and PEP' the hour circle of a star. The right ascension (α) of this star is the angular distance between the hour circle of the origin (which passes through the First Point of Aries γ) and the hour circle of this star E. Let aOE be the angle between this star and the celestial equator. We call this angle the declination (δ) of the star E.*

and arc seconds ("). It is negative for a star in the austral hemisphere and positive for a star in the boreal hemisphere. **Right ascension** is the second coordinate we need to locate a star on the celestial sphere, and is equivalent to terrestrial longitude. It is also called the alpha coordinate, or α. The right ascension of the star E is located on its **hour circle**, which is equivalent to the terrestrial meridian and is represented by the great circle joining the poles and perpendicular to the celestial equator. Just as there is a meridian of origin (the Greenwich meridian), there is an hour circle of origin, located at the intersection of the ecliptic with the celestial equator. Astronomers call this point the **First point of Aries**, or the γ (gamma) point, which indicates

the position of the Sun on the first day of spring (Fig. 6). Right ascension is measured in units of hours (h), minutes (m) and seconds (s), and is called **sidereal time**.

We have thus defined the two coordinates necessary to locate a star in the sky. Because the perturbation forces of the Sun and the Moon cause various oscillations and irregularities in the motion of the Earth, these coordinates are not fixed in time. The First Point of Aries, the origin of right ascension, shifts each year on the celestial equator. Moreover, stars have a motion of their own which, despite being a very slow and very small movement, does change their position as time passes. Astronomers therefore have to revise their stellar catalogues periodically and adopt a new reference coordinate system, which itself will become obsolete in a few years and will have to be changed again. The art of astronomy is one of meticulousness and patience ...

## Magnitude

Our ancestors talked about "big" or "small" stars, not because of their size but because of their brightness. Ptolemy (*c.* AD 90–168), a Greek

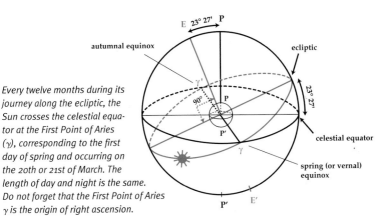

*Every twelve months during its journey along the ecliptic, the Sun crosses the celestial equator at the First Point of Aries (γ), corresponding to the first day of spring and occurring on the 20th or 21st of March. The length of day and night is the same. Do not forget that the First Point of Aries γ is the origin of right ascension.*

| CONVERSION OF COORDINATES | |
|---|---|
| Declination | Right ascension |
| 360° | 24 h |
| 15° | 1 h |
| 15′ | 1 min |
| 15″ | 1 s |

astronomer, mathematician and geographer, was the first to attempt a classification. In his catalogue he divided 1028 stars into six groups according to their magnitude. The most brilliant stars were of the first magnitude, while the last stars to be visible with the naked eye were of the sixth magnitude. The arrival of instruments that could measure the intensity of light towards the end of the last century and the compilation of the first large stellar catalog led astronomers to develop a more pre-

cise system of measurement. Norman Pogson introduced the modern definition of **magnitude**, which is defined by the formula:

$$m = -2.5 \log E$$

where $m$ is the magnitude and $E$ is the brightness of the star in units of lux. The ratio of brightness between two stars whose magnitudes differ by one is 2.512. Therefore, a star of first magnitude is 2.512 times brighter than a star of second magnitude, which is itself 2.512 times brighter than a star of third magnitude, etc... Between a star of first magnitude and a star of sixth magnitude (the last stars visible with the naked eye), the ratio of brightness is

$$2.512^5 = 100.$$

We are talking here about the **apparent visual magnitude**. Several

| | Limit of magnitude | Visible objects | |
| --- | --- | --- | --- |
| | | Stars | Galaxies |
| Naked eye | 6 | 6 000 | 3 |
| ∅ 50-mm binoculars | 10 | 300 000 | 300 |
| ∅ 60-mm refractor | 11 | 600 000 | 1 000 |
| ∅ 115-mm reflector | 12.4 | 2 300 000 | 1 800 |
| ∅ 200-mm reflector | 13.6 | 5 700 000 | 4 000 |
| ∅ 400-mm reflector | 15.1 | 30 000 000 | 10 000 |

∅ is the diameter of the instrument

The luminosity ratio $E_1/E_2$ between a star of apparent visual magnitude (mv) equal to $m_1$ and one of apparent visual magnitude $m_2$ can be expressed as: $E_1/E_2 = 10^X$, with $X = \frac{2}{5}(m_2 - m_1)$.

### SOME APPARENT VISUAL MAGNITUDES

| Object | Magnitude | Remarks |
| --- | --- | --- |
| Sun | −27 | |
| Moon | −12 | Full Moon |
| Venus | −4.1 | At greatest elongation |
| Jupiter | −2.4 | At opposition |
| Sirius | −1.6 | The brightest star |
| Arcturus | 0 | |
| Polaris | +2.1 | The Pole Star |
| 1 square degree of dark sky | +3.5 | Average value, at the zenith and with no Moon |
| M31 | +4.8 | Andromeda Galaxy |
| Pluto | +14.7 | At opposition |

symbols can be used to represent it: "mv" or "m" or "mag". It is:

- a "visual" magnitude because the eye is the receptor for the light coming from the stars,
- an "apparent" magnitude because we can only see the apparent brightness of the stars and not their intrinsic brightness.

Catalogues frequently add the **photographic magnitude** of objects, "*mp*", or sometimes "$m_{pg}$", which is defined by the standardized **UBV system** (near-ultraviolet, blue and visual). Astronomers use the UBV system to measure brightness and hence magnitudes for a specific wavelength, and then to determine

The **Parsec** is a unit of measure, which corresponds to the distance of a star whose annual parallax would be of 1°. One parsec equals 3.26 light years, i.e. 206 265 astronomical units. The **astronomical unit** (AU or a.u.) is the mean distance between the Earth and the Sun, i.e. just under 150 billion kilometers.

### APPARENT SIZE OF SOME OBJECTS

| | |
|---|---|
| Andromeda Galaxy | $2.5° \times 0.7°$ |
| Orion Nebula | $1.1° \times 1°$ |
| Moon | $0.5°$ |
| Sun | $0.5°$ |
| Jupiter | $30''$ to $50''$ |
| Saturn | $15''$ to $20''$ |
| Venus | $6''$ to $60''$ |
| Uranus | $4''$ |

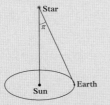

*The annual parallax $\pi$ is the angle between the Sun and the Earth, as seen from the star.*

the star's **color index**. These color indices are very important for our understanding of stellar populations, since they give information about the energy emitted by an object at a certain stage in its evolution.

The **absolute magnitude** $M$ denotes the apparent brightness that stars would have, were they all at the same distance from the Earth, a distance arbitrarily fixed at 10 parsecs.

Apart from a few rare exceptions, in particular for the observation of variable stars, amateur astronomers do not often use the UBV system and absolute magnitude. We shall therefore only use the apparent visual magnitude in the rest of the book.

## Apparent size

In this first chapter we have defined two things about each celestial object that we shall look at. The first is the position of this object on the celestial sphere, determined by the intersection of two precise coordinates: declination and right ascension. The second is the luminous quantity of energy, measured from the Earth on a scale of magnitude, that this object emits or reflects. Depending on what we want to find out about the object, we can study its apparent visual magnitude, its UBV magnitude, or its absolute magnitude.

In addition to position and magnitude, the amateur astronomer needs a third measurement for the object he wishes to observe: its **apparent size**. The apparent size of a star is expressed in units of degrees (°), arc minutes (') and arc seconds ("). It is not only dependent on the object's distance, but is also a function of its absolute size.

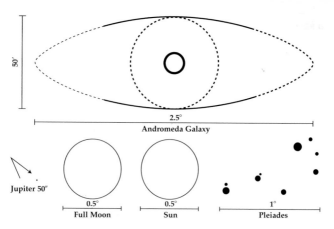

50'

2.5°
**Andromeda Galaxy**

Jupiter 50"

0.5°
**Full Moon**

0.5°
**Sun**

1°
**Pleiades**

*Comparison between the apparent diameters of various objects.*

The notion of apparent size is very important for the amateur astronomer, since it is related to the **resolving power** (or power of resolution) of the telescope or binoculars.

If we assume that our observational instrument enables us to distinguish between two stars separated by at most 2 arc seconds, an object whose apparent diameter is 1 arc second cannot be resolved, even in the best seeing conditions.

Table opposite: the resolving power (or resolution) of an instrument is its ability to perceive details, to separate double stars for example. The resolving power of a lens or of a mirror is related to its diameter. It can easily be calculated with the formula:

$Ps = 12/D$ (in good seeing conditions, without turbulence)

or

$Ps = 30/D$ (when observational conditions are affected by atmospheric turbulence).

Where $D$ is the diameter of the lens or the mirror, in centimeters, and $Ps$ is the resolving power in arc seconds).

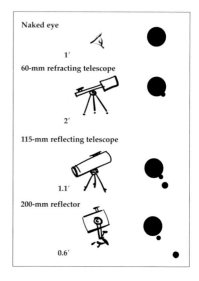

Naked eye

1'

60-mm refracting telescope

2'

115-mm reflecting telescope

1.1'

200-mm reflector

0.6'

## HOW TO READ A TYPICAL ENTRY

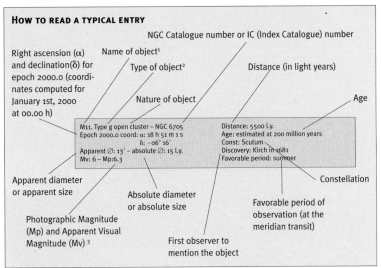

NGC Catalogue number or IC (Index Catalogue) number

Right ascension (α) and declination(δ) for epoch 2000.0 (coordinates computed for January 1st, 2000 at 00.00 h)

Name of object[1]

Type of object[2]

Nature of object

Distance (in light years)

Age

M11. Type g open cluster – NGC 6705
Epoch 2000.0 coord: α: 18 h 51 m 1 s
δ: –06° 16′
Apparent ∅: 13′ – absolute ∅: 15 l.y.
Mv: 6 – Mp:6.3

Distance: 5500 l.y.
Age: estimated at 200 million years
Const: Scutum
Discovery: Kirch in 1681
Favorable period: summer

Apparent diameter or apparent size

Absolute diameter or absolute size

Constellation

Favorable period of observation (at the meridian transit)

Photographic Magnitude (Mp) and Apparent Visual Magnitude (Mv) [3]

First observer to mention the object

[1] The name of an object is often identified by its number in the Messier Catalogue (M) or in the New General Catalogue (NGC).

[2] **Open cluster** types (using the Shapley classification):
  c: very open and irregular cluster
  d: open and poor cluster
  e: moderately rich cluster
  f: rich cluster
  g: very rich and dense cluster
- **Globular cluster** types: The Shapley classification adopted here distinguishes clusters I to XII in order of decreasing concentration.

- **Galaxy** types: The Hubble classification distinguishes five large families of galaxies.
  elliptical galaxies: from E0 to E7,
  lenticular galaxies: So,
  normal spirals: Sa, Sb, Sc, …
  barred spirals: SBa, SBb, SB
  irregular galaxies: Ir I, Ir II, …

[3] For planetary nebulae, the Mv entry followed by an asterisk indicates the visual magnitude of the dwarf star located at its center.

## THE HUBBLE CLASSIFICATION TABLE

The subdivisions given here in the groups of galaxies are a function of the openness of the arms of the galaxies

| Elliptical and lenticular galaxies | Normal spirals | Barred spirals | Irregular galaxies |
|---|---|---|---|
| E0 | Sa — Tightly closed arms | SBa | |
| E3 | Sb — Open arms | SBb | Ir |
| E7 | Sc — Very open arms | SBc | |
| So | | | |

# THE SOLAR SYSTEM

# THE SUN

Located at an average distance of 149 600 000 km from the Earth, the Sun is the star we know best. Its observation, however, can present a real danger for the astronomer and some elementary precautions are essential.

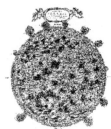

*Engraving from 1635 representing the Sun* (author's collection).

## Precautions

Never look at the Sun with the naked eye without a filter, since its brightness can cause irreparable damage to the eye. Observed with a filter, or with the naked eye at sunrise or sunset, the Sun is a sphere whose apparent diameter varies between 31′32″ and 32′36″.

There are two safe ways to look at the Sun. The first solution is to project its image onto a white screen placed behind the eyepiece. Take care though not to use orthoscopic or Plössl eyepieces whose elements are cemented, since at high temperature the adhesive melts and spreads across the lenses of the eyepiece. It is better therefore to choose the "H" or "HM" series in which the lenses are not cemented. For 60- to 150-mm

instruments, the original 20-mm H eyepiece is perfect for these observations by projection. Finally, do not attempt solar projection with a Schmidt–Cassegrain type telescope, because this would damage

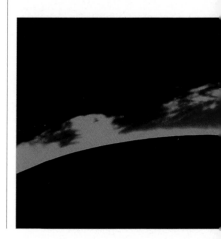

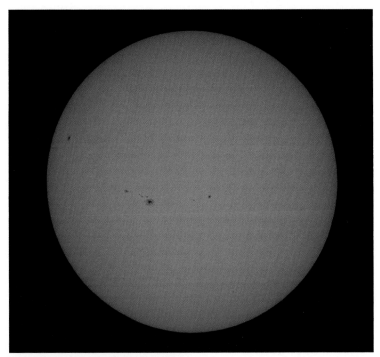

*Using an aperture of at least 50 mm and a filter, we can study the evolution of sunspots, groups of sunspots and faculae on the solar surface. This photograph was taken with a 106-mm fluorite refractor, set up as shown on page 16.*

*The use of a coronagraph creates an artificial eclipse. The observation of solar protuberances then reveals the amazing activity of our star.*

the coatings of the Schmidt plate or would unglue the secondary mirror.

The second solution, and the safest, is to place over the instrument's aperture a semi-aluminized filter whose transmission is only one ten thousandth (1/10 000). The observed image is still bright enough to allow the easy study of spots, penumbrae and faculae. Using this second method, there are no time limits and it is completely safe.

All those wishing to embark on a thorough and regular study of the Sun are advised to invest in such a filter, whose cost is modest compared to the guarantee of safety it provides (about $125 (£80) for a 115-mm reflector).

Whatever solution you choose, do not forget to block the aperture of your finder in order to protect it, otherwise the reticle might be damaged beyond repair.

## Observing and counting sunspots

The Sun is a **variable star**, as shown by the nature and evolution of sunspots. Sunspots are regions of the Sun's surface cooler than the photosphere and thus appear darker.

*An amateur's equipment set up for solar astrophotography. In front of the refractor, the semi-aluminized filter only transmits 1/10 000 of the brightness, but allows the whole instrument aperture to be used. Note that the camera attached to the refractor is mounted on a tripod to minimize any vibration caused by the camera mirror during the exposure.*

Strong magnetic fields in the spots area are the cause of this cooling effect.

One of the main activities for amateurs is to note down and study the evolution of these spots on the solar surface, in order to throw light on the 11-year cycle of activity of our star.

*The Hα filter allows the observation of all solar activity in the hydrogen spectral band.*

Sunspots can be of many different sizes, appearances and durations. The penumbrae surrounding them form a grayish band of aligned streaks.

Faculae, more visible at the darkened edge of the solar disk, are bright patches near groups of sunspots. They are recognizable by their white color and spreading forms. Sunspot groups often originate within them.

Solar activity is measured by the Wolf number, given by the following formula :

$$W = R = k(10g + t)$$

where $g$ is the number of spots or groups of spots, $t$ is the total number of spots and $k$ is a coefficient close to 1 depending on the observer, his/her instrument, and the observational conditions.

# THE MOON

The Moon, Earth's natural satellite, is one of the biggest moons of the solar system (3436 km in diameter). While professional astronomers have tended to neglect it since the end of the lunar missions of the 1970s, our closest neighbor has remained the favorite object of amateurs. The knowledgeable observer will recognize the Moon's formations and will always rediscover them with the same pleasure, and some astronomers have specialized in the careful study of its surface.

Because of its constant movement and the cycle of its phases, which takes just under a month, the landscape of the Moon changes spectacularly. Many observers report transient phenomena such as sudden brightening, changes in color or detail, etc., showing that we still do not know everything about our satellite.

## Observation

The Moon may be observed with anything from the smallest of instruments, and binoculars alone

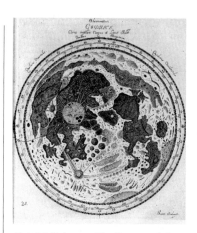

*First detailed map of the Moon, made by the German astronomer J. Hevelius in 1647* (Bibliothèque Nationale, Paris. © Bibl. Nat./ Archives Photeb).

can reveal countless moonscapes. Seas appear and the main basins and craters show their complex structure. Using a 60-mm refractor with a magnification of 60 to 90 times, the observer will see a multitude of geographical variations: mountain chains and unequal areas on the lunar surface cast their shadows. Regions close to the **terminator** are particularly interesting since the grazing sunlight offers unforgettable sights.

*This photograph, taken with a 106-mm refractor, shows the lunar landscape as seen with binoculars or a low magnification reflector*
(© C. Lehénaff).

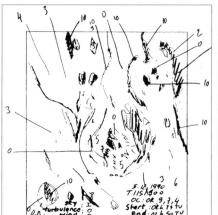

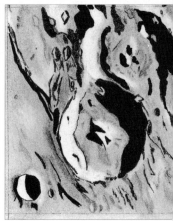

*Left: To sketch a drawing such as this through the eyepiece might take ten to twenty minutes. The observer has labelled the gray tones from 0 (for white) to 10 (for the dark background sky).*

*Right: A refinement of the previous sketch. This drawing, using a T 115/900 reflector (Or 9-mm, 7-mm and 4-mm eyepieces), shows the Cyrillus (diameter of 90 km), Theophilus (100 km) and Mädler (32 km) basins, from north to south.*
(Drawings: H. Burillier, December 1990.)

When the atmosphere is still, take advantage of the low turbulence to use a higher magnification. With a magnification twice the diameter of the instrument, you can study the microcraters which cover larger craters and follow for many kilometers the sharp crack-like clefts (or rills) which run between cirques and hills.

The Moon is an excellent subject for training the eye. With some experience you will learn to distinguish more details, some of which will be fleeting, appearing and disappearing due to atmospheric turbulence, while vast plains will change color according to the lighting.

Astronomical drawing is of particular value here. Train yourself to draw a crater, a shadow, a basin. With a bit of practice, you will soon be able to distinguish so many different things in a lunar formation that it will be impossible to draw everything you see.

*The lunar disk a few hours before full Moon. Photograph taken using a 106/900 fluorite refractor with a 1.8 tele-extender* (© C. Lehénaff).

Using wide and ultra wide field eyepieces is great for lunar observations. An investment in polarizing filters is really worthwhile: they will not only reduce the overall brightness around the time of full Moon, but will also noticeably enhance the tonal variations of lunar structures.

# MERCURY AND VENUS

## Mercury

Mercury is the nearest planet to the Sun, travelling around it at an average distance of 57 910 000 km, with a maximum **elongation** of 28°. Slightly larger than the Moon, this small body is particularly difficult to locate at twilight or just before sunrise. Mercury reaches its maximum brightness at superior **conjunction**, and then has a magnitude of −2, greater than that of the most brilliant star, Sirius.

Using a telescope to view Mercury is of little interest, and only the changing phases deserve attention. The apparent diameter of Mercury's disk varies between 12.9″ and 4.7″. A 1″ detail observed from the Earth represents more than 670 km on the planet.

## Venus

There are many observers who have never seen Mercury; Copernicus is a famous example. On the other hand, it is unlikely that there is anyone who has never seen the shining light of Venus. Also known as the evening star, Venus reaches a magnitude of −4.4 (12 times the brightness of Sirius)

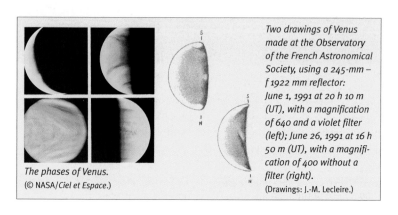

The phases of Venus.
(© NASA/*Ciel et Espace*.)

*Two drawings of Venus made at the Observatory of the French Astronomical Society, using a 245-mm – f 1922 mm reflector: June 1, 1991 at 20 h 10 m (UT), with a magnification of 640 and a violet filter (left); June 26, 1991 at 16 h 50 m (UT), with a magnification of 400 without a filter (right).*
(Drawings: J.-M. Lecleire.)

and can be located in broad daylight with amateur instruments.

Venus circles the Sun at 108 210 000 km on average, along an almost circular ellipse. It moves from one side of the Sun to the other, up to 48° at maximum elongation.

## Observation of Venus

Using a 60-mm refractor or a more powerful instrument with a magnification slightly greater than 100, the amateur can follow the changing phases of Venus. A 115/900 reflector with a 7-mm eyepiece will sometimes reveal a few details in the thick clouds of its atmosphere. A blue filter (80A) will facilitate the observation of this particularly bright object, diminishing the scintillation and thus allowing an easier observation of the crescent or the globe.

Nonetheless, Venus remains a mysterious object covered by a thick hazy layer, which prevents any observation of its surface. The observer sees little more than a whitish area with occasional irregular flecks near the **terminator**, due to diffusion of sunlight in the high atmosphere of the planet.

*Mercury's surface looks very much like that of the Moon. However, no detail can be seen from the Earth, and the observer has to be content with observing the evolution of the planet's phases.*

At maximum elongation – at a quarter phase – 1″ observed from the Earth represents 513 km at Venus's surface.

# MARS

Mars is the most Earth-like planet, after Venus. It has a very elliptical orbit around the Sun, bringing the planet to within about 56 000 000 km from the Earth at **opposition**, which is the most interesting time for observation, and which is found in astronomical ephemeris.

## Observation of Mars

The observer may study the surface of the Red Planet with any instrument. Mars is so bright (magnitude −2.8) that finding it is very easy, since its bright red color differentiates it from the stars. Mars rotates slower than the Earth and completes its day in 24 h 37 m 22.6 s. The observer will see each night almost the same region of the martian surface, which shifts slightly from night to night, and will be able to make a complete map of the planet in just over one and a half months.

The evolution of the white polar cap can be studied using an 80-mm refractor with a light green filter (No. 56). The sudden appearance of martian storms and their violent landscape modifications are the subject of many discoveries, and the use of a red–orange filter is strongly advised (No. 21 and/or 23A).

Astronomers wishing to study Mars seriously need to orientate their drawing, by using a suitable grid of parallels and meridians, and geographic and/or celestial north and south must be precisely located. The prime meridian on Mars passes along Sinus Meridiani, whose position is given in astronomical ephemeris at oo h UT. Each drawing must indicate the longitude of the central meridian ω (omega) at the time of observation. Remember that the central meridian must be given for the martian geometric disk, and not just for the visible part of the disk, at the time of observation (see the example opposite).

*Mars, January 18, 1995. The planet was at 0.7 a.u. from the Earth, with an apparent diameter of 13″ (Mv 0.7. T C8, magnification of 200 and 330, light blue and orange filters).*
(Drawing: H.Burillier.)

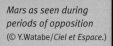

*Mars as seen during
periods of opposition*
(© Y.Watabe/*Ciel et Espace.*)

Calculation of the central meridian of Mars for an observation on December 30,
1994 at 22 h 35 m.
For December 30, 1994 at 00 h UT, ephemeris gives $\omega = 160° \, 15'$.
Then:

| | |
|---|---|
| Longitude of the central meridian at 00 h UT | $160° \, 15'$ |
| Rotation in 22 hours | $+321° \, 65'$ |
| Rotation in 30 minutes | $+7° \, 31'$ |
| Rotation in 5 minutes | $+1° \, 21'$ |
| | $490° \, 32'$ |
| | $-360°$ |
| Longitude of the central meridian at 22 h 35 m | $130° \, 32'$ |

# JUPITER

Jupiter is the king of the giant gas planets. It is the biggest planet of the solar system, and orbits around the Sun in almost 12 years. Its apparent diameter varies between 30″ and 50″ depending on whether it is at **conjunction** or **opposition** with the Sun. The amateur astronomer can observe Jupiter at any time in its orbit, and see many features.

## Observation

The well-known dark tropical belts are clearly visible using any instru-

The impact of the comet SL9 on Jupiter was one of the major events of the end of our century. The major impact sites (the brown patches on this photograph from the Hubble Space Telescope) have been observed using small diameter amateur instruments in summer 1994.

(© NASA/*Ciel et Espace*.)

| JJ: 244 | Date: / / | No: |
|---|---|---|

S.A.F. Nov. 1987

R. Caron

| | | Hours UT |
| | | SSTB: |
| | | STB: |
| | | SEB: |
| | | EZ: |
| | | NEB: |
| | | NTB: |
| Notes: ............ ᴺ | Start: .......UT | NNTB: |
| ........................ | Finish ......UT | |

| Observer: | Instrument | .D | .F/D |
|---|---|---|---|
| Place: | Magnification | | Filter |
| Latitude: | Zone | UT | CM |
| Longitude: | | | |
| Altitude: | | | |
| Wind strength: | | | |
| Direction: | | | |
| Estimates: Transparency: | | | |
| Temperature: | | | |
| Turbulence: | | | |
| Quality of the observation: | | | |

*Template for Jupiter produced by the Commission for the Planets.*

(© French Astronomical Society.)

| Code D | Dark features | Description of dark features |
|--------|---------------|------------------------------|
| SDER | | **Spot**, small and very dark, surrounded by a bright annulus |
| COL | | **Column** – Dark column-shaped region which may be perpendicular to the zone or at a slight angle. |
| SPOT | | **Spot** – Discrete unstretched spot, without a dark surrounding annulus |
| DIST | | **Perturbation** – Large dark or blackish region, more or less well defined, usually speckled with finer details of unusual form. |
| SECT | | **Section** – A darker area within a band or a zone |
| FEST | | **Festoon** – Festoon-like filaments, crossing or leaving a zone, in a buckle-like form. One or both ends of a festoon might lead to a dark condensation or a groove on the border of a belt. |
| STRK | | **Stripe** – Very dark and elongated stripe |
| PROJ | | Protuberance on the edge of a belt, which may or may not be darker than the main part of the belt. This kind of object varies from a small round bump to a spike-like excrescence |
| BAR | | **Bar** – Dark elongated area |
| VEIL | | **Veil** – Large dark uniform region sometimes found in the polar regions and temperate zones. |
| P.BELT | | Preceding **edge** of a dark patch or longitudinally stretched region, a few degrees wide. |
| F.BELT | | Following **edge** of a dark patch or longitudinally stretched region, a few degrees wide. |

ment including 11×80 binoculars mounted on a tripod, and a 115/900 reflector allows the study of the jovian poles at any time.

These are yellow, sometimes slightly grayish, in color. The tropical regions display many peculiarities in the gaseous structures of the

## NOMENCLATURE OF VISIBLE DETAILS IN THE JOVIAN ATMOSPHERE

| Code W | Light features | Descriptions of light features |
|---|---|---|
| AREA | | **Surface** – Large bright region, irregular in shape or diffuse |
| BAY | | **Bay** – Large indentation, usually a half-oval in a belt's border |
| GAP | | **Gap** – Rather large weak or missing area in a belt |
| NICK | | **Nick** – Small semi-circular indentation in the edge of a belt usually slightly brighter than the adjacent zone |
| SPOT | | Very bright **spot**, usually circular |
| OVAL | | **Oval** – Large or medium-sized oval region, rather bright and well defined |
| SPTR | | **Spot** – Small and very brilliant spot surrounded by a dark ring |
| SECT | | **Section** – Brighter part of a belt or a zone |
| STRK | | **Ray** – Very elongated bright spot. When such a formation appears on a belt, it can look like a fragment of a fissure |
| RIFT | | **Rift, split** – The rift looks like an irregular break in a band. |
| P.BELT | | Preceding **edge** of a light spot or longitudinally stretched region, a few degrees wide. |
| F.BELT | | Following **edge** of a light spot or longitudinally stretched region, a few degrees wide. |

(© French Astronomical Society)

high atmosphere, and a trained eye will be able to distinguish the dark lumpy areas which attenuate in twisted filaments and seem to link belts together.

The Great Red Spot is visible using a 60-mm refractor, and the use of a blue filter (No. 38A) will enhance the contrast for its observation and for that of other cloudy structures of Jupiter.

As for Mars, drawings of Jupiter must note the longitude of the central meridian, ω. But since Jupiter rotates very rapidly, two reference systems must be adopted: the first

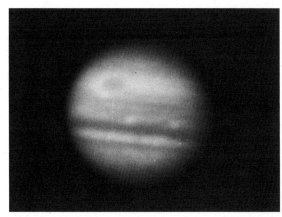

*Jupiter photographed using a 245-mm – f 1922 mm reflector and exposure time of 2 s, from the Observatory of the French Astronomical Society, in central Paris.*
(© J.-M. Lecleire.)

system corresponds to the equatorial zone, of sidereal rotation period of 9 h 55 m 30 s, while the second reference system corresponds to the latitude of the Great Red Spot, which has a period of rotation of 9 h 55 m 41 s. These data are given in astronomical ephemeris.

Jupiter is a fascinating world, constantly changing and frequently subject to turbulence; the evolution of violent storms makes it a particularly interesting study. The dance of the four main satellites is also very striking. Io (1.1″), Europa (0.9″), Ganymede (1.6″) and Callisto (1.45″) orbit the jovian disk, sometimes occulted, sometimes projecting a tiny shadow when they pass between Jupiter and the observer.

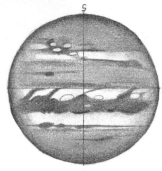

*A drawing of Jupiter, March 14, 1991, at 18 h 55 m (UT), from the Observatory of the French Astronomical Society, Sorbonne, Paris, using the 153-mm – f 2300 mm refractor and a magnification of 280.*
(Drawing: J.-M. Lecleire.)

# Saturn

Using a refractor, Saturn must surely be the most beautiful object in the sky, and once seen it will never be forgotten.

## Observation

The last planet of the solar system which can be seen with the naked eye, Saturn reaches a maximum magnitude of −0.4, slightly brighter than Vega. Using binoculars on a tripod, the amateur can make out Saturn's form, elongated because of its rings. Yellowish in color, its apparent diameter is, however, rather small, 20″ at most. Adding the rings at each side of the disk, the apparent size of the planet is almost that of Jupiter.

Using a 60-mm refractor with a ×60 magnification, the astronomer can make worthwhile observations. The rings are then well defined, and it becomes possible to distinguish the Cassini division, a dark gap between the two main rings. Titan, the largest satellite, is visible like a small star of 8.3 magnitude. The Cassini

division is perfectly visible using a 115/900 reflector with a 6-mm Or/HD eyepiece (magnification of 150). At the maximum openness of the ring system, the observer can study the shadow of the rings on the disk of the planet or, alternatively, Saturn's own shadow on the rings.

A grayish band is also visible on the disk next to the tropical regions, which are sometimes subject to intense perturbations and which will provide the amateur observer with a remarkable opportunity for investigation. The observation and evaluation of the bright intensities the perturbations produce must be noted down

*Saturn seen with a 70/480 Pronto refractor, from a sixth-floor balcony in Paris, March 15, 1994, at oo h 35 m (UT), using a 8.8-mm UWA eyepiece (×54 magnification), a Barlow lens ×2 and a yellow filter (No.12).* (Drawing: H. Burillier.)

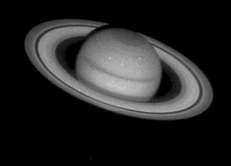

*Saturn seen with the Hubble Space telescope* (© NASA/*Ciel et Espace*).

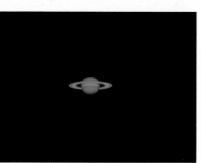

*Saturn as seen with an amateur astronomer's refractor or reflector* (© S.Brunier/*Ciel et Espace*).

meticulously; a light green (No. 56) and/or an orange filter (No. 21) can enhance these details.

Other satellites are visible apart from Titan: Janus (mv 10), Rhea (mv 10) and Tethys (mv 10.5). Dione (mv 10.7) is more difficult to see. Using an aperture of at least 150 mm and under good seeing conditions, the Encke division can be made out within the ring system.

# OPEN CLUSTERS

# χ AND λ PERSEI

## History

Half way in the Galaxy between α Persei and δ Cassiopeiae stars, NGC 869 and NGC 884 form the famous pair of open clusters of Perseus, known as the Sword Handle, or the Double Cluster in Perseus. Well known since antiquity – Hipparchus mentions it in 150 BC – this splendid stellar body is one of the treasures of the autumnal sky. With the naked eye, the attentive astronomer should be able to separate this double cluster, which looks like a circular, milky cloud. Both structures have an apparent diameter close to that of the full Moon, half a degree, and are separated from each other by the same amount.

## Observation

Binoculars will probably reveal the most striking image. Twenty or so very bright stars (mv 6 and 7) can be seen with 7×50 binoculars, while two bright archipelagos of stars fill the whole field of 11×80 binoculars. This swarm of over 100 stars is a magical sight.

*χ and λ Persei (© A.Fujii/Ciel et Espace).*

*χ Persei (or NGC 884) and λ Persei (or NGC 869) (© J.-C. Merlin/Surelées Observatory).*

*The field of these photographs is of 20"/mm, i.e. 30' with a 40-cm reflector with an image intensifier. It more or less corresponds to the field obtained with a 9-mm Or/HD eyepiece on a 115/900 reflector.*

λ Persei
Type f open cluster
NGC 869
Epoch 2000.0 coord: α: 02 h 19 m 0 s
                    δ: +57° 09'
Apparent ⌀: 30' – Absolute ⌀: 100 l.y.

Mv: 4.1 – Mp: 4.4
Distance: 7000 to 8000 l.y.
Age: 12 million years
Const.: Perseus
Favorable period: autumn, winter

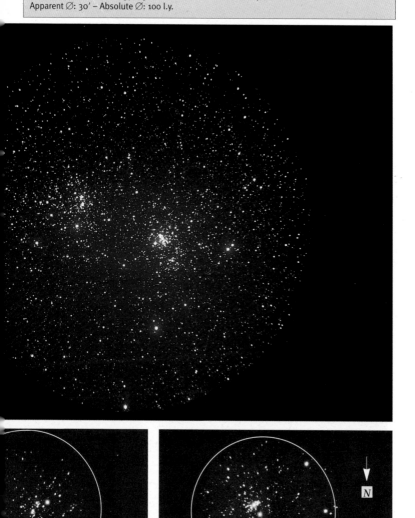

NGC 869 contains 400 stars up to twelfth magnitude, while NGC 884 contains 300 stars, also up to magnitude 12.

The Double Cluster in Perseus is made up of blue, very hot stars, or **supergiants**, whose luminosity is equivalent to 5000 Suns. Within this very young formation, a few very massive stars, much older, have exited the **main sequence** and have become **red giants**. Most of them, of magnitude between 7.6 and 9.2, are clearly visible with amateur instruments, and are distinguished by their red color.

## Other observations

Make sure you look for NGC 957 in this region of the sky. This small open cluster, 1°30′ east–northeast of χ and λ Persei, contains 40 stars of magnitude 11 to 15, and has a diameter of 10′. It also contains two close binary stars, of which one is visible with a 60-mm refractor: h 2143. They are of magnitude 8 and 8.5 and are separated by 23.4″.

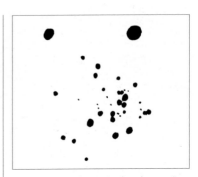

*Central region of NGC 884 (200/2000 reflector, ×230, field of 22′)* (Drawing: H.Burillier).

NGC 957
Type e open cluster
NGC 869
Epoch 2000.0 coord: α: 02 h 33 m 6 s
δ:+57° 32′
Apparent ⌀: 10′ – Absolute ⌀: 12 l.y.?
Mv: 7.1
Mp: 7.6
Distance: 7000 l.y.
Constellation: Perseus
Favorable period: autumn, winter

*The Double Cluster in Perseus*
*(field: see p. 34)*
(© J.-C. Merlin/Surelées Observatory).

χ Persei
Type e open cluster
NGC 884
Epoch 2000.0 coord: α: 02h 22m 4s
δ: +57°07'
Apparent ∅: 30' – Absolute ∅: 100 l.y.

Mv: 3.9 – Mp: 4.7
Distance: 7000 to 8000 l.y.
Age: 12 million years
Const.: Perseus
Favorable period: autumn, winter

# THE HYADES

The Hyades Galactic Cluster is the closest of all to us, and in consequence is the most spread out, since it covers almost 6° of the sky. All star types are present in this vast formation, including **giants** and **white dwarfs**.

*In the constellation Taurus, the Pleiades and the Hyades are two young star clusters, both visible with the naked eye* (© H. Burillier).

Aldebaran, a splendid red and luminous star (mv 0.8) is the most conspicuous object of the Hyades group, although in fact it is not related to it.

## Observation

The size of this cluster means that it can only be seen in its totality with the naked eye, and in good seeing

Pleiades

Hyades

| The Hyades (Mel 25) | Mv: 0.8 – Mp: 0.8 |
| Type c open cluster | Distance: 130 l.y. |
| Epoch 2000.0 coord: α: 04 h 16 m 7 s | Const.: Taurus |
| δ: +15° 31′ | Favorable period: winter, the whole night |
| Apparent ⌀: 6.7° – Absolute ⌀: 18 l.y. | |

conditions. In the field of a 115/900 reflector with a low magnification eyepiece (AH 40-mm: 22.5×), the cluster will seem very spread out and only a few scattered stars will be seen.

Two hundred and sixty stars can be seen, up to 9th magnitude, over such a large area that it would take twelve full Moons to cover it!

*A vast open cluster with the reddish Aldebaran the most conspicuous star, the Hyades are visible throughout the winter in the first half of the night* (© A. Fujii/*Ciel et Espace*).

Generally speaking, observers should remember that great magnifications will be of little use in studying the Hyades, and should content themselves with the naked eye and binoculars.

# M11

Other name **WILD DUCK CLUSTER**

## Search

M11 is an exceptional galactic cluster located right in the middle of the Milky Way, in the arm of the Scutum. Easily visible with the naked eye when conditions are favorable, M11 is without doubt the most beautiful open cluster of the boreal sky.

## History

Gottfried Kirch, of the Berlin Observatory, discovered it in 1681. Its star density is so high (estimated at 83 stars per cubic parsec) that controversy about its exact nature lasted for many years, and for a long time it was considered to have been a globular cluster, until spectroscopic measurements definitely ranked it in the open cluster category.

## Observation

Right from the lowest magnifications, refractor or reflector observations show a considerable agglomeration of stars. Sixty or so stars are visible with a 60-mm refractor and a magnification of 56, over 100 can be seen with a 200-mm reflector, while 600 can be observed up to

15th magnitude. It is estimated that M11 contains over a thousand members ... Its image, dense and compact, is magical.

Contrary to most other open clusters, M11 can be observed at higher magnifications (1.5 times the instrument diameter). The use of a short

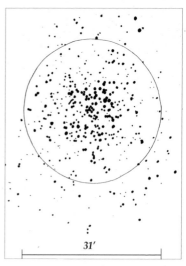

31'

*M11 seen in a 115/900 reflector, Or/HD 9 and 7-mm eyepieces, with a magnification of 100 and 128, fields of 31' and 24'* (Drawing: H. Burillier).

M11. Type g open cluster – NGC 6705
Epoch 2000.0 coord: α: 18 h 51 m 1 s
δ: −06° 16′
Apparent ∅: 13′ – absolute ∅: 15 l.y.
Mv: 6 – Mp: 6.3

Distance: 5500 l.y.
Age: estimated at 200 million years
Const.: Scutum
Discovery: Kirch in 1681
Favorable period: summer

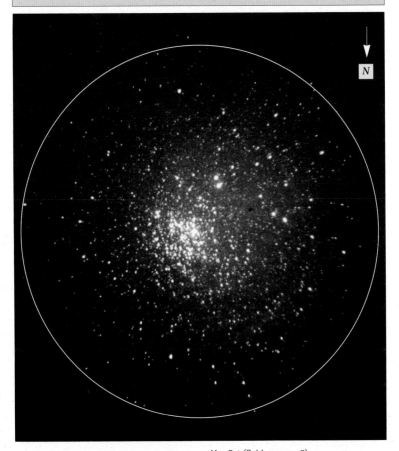

N

*M11 Sct (field: see p. 48)*
(© J.C.Merlin/Surelées Observatory).

focal length eyepiece allows the
central cluster luminosity to be
reduced, and this region then looks
similar to a sphere made up of an
infinite number of spikes of light.

# M23

Oddly enough, many astronomers neglect this beautiful cluster, probably because this rich region, where the constellation Sagittarius lies, contains numerous splendid nebulae. M23 lies to the north of Sagittarius, just at the edge of the Milky Way and just above the background sky limit, 4° north of the bright Trifid Nebula.

Messier was the first to note its existence on the 20th of June 1764, and John Herschel, despite the mediocrity of the instruments of his time, counted 100 or so stars of magnitudes between 9 and 13 in M23.

## Search

Easy to locate in the segment made by ξ and μ Sagittari, M23 is a splendid concentration of white and blue stars, whose diffuse and bright aspect is clearly revealed with a 6×30 finder.

## Observation

Just visible to the naked eye, a star located 20' northeast of the cluster determines the position of the group; the large number and variety of stars in this region does not make the astronomer's job very easy! Stars located at the edge of M23 can be isolated with 11×80 binoculars, while a 115/900 reflector, with a magnification of 72 (Or/HD 12.5-mm eyepiece), will reveal a hundred or so stars up to mv 9. A striking feature is an odd, almost perfect arc of stars. As with most open clusters, M23 is made up of blue, very hot stars (spectral type B); some **giants** have also been observed.

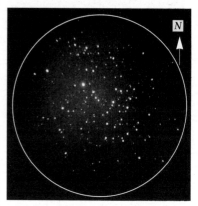

*M23 Sgr (field: see p. 34)* (© J.-C. Merlin/ Surelées Observatory).

M23. Type e open cluster
NGC 6494
Epoch 2000.0 coord: α: 17 h 56 m 8 s
                    δ: −19° 01′
Apparent ∅: 27′ – absolute ∅: 15 l.y.

Mv: 5.5 – Mp:6.9 – Distance
Const.: Sagittarius
Discovery: Messier in 1764
Favorable period: June and
of the night

*M17*

*M18*

*M24*

*M25*

μ

*M22*

λ

*The Milky Way with the stellar cluster M23. This photograph, using a 135-mm telephoto lens with an exposure time of 15 min (on T Max 400) corresponds fairly well to what one would see with simple binoculars in this region of the sky under very good seeing conditions.* (© Jean-Marc Lecleire.)

# M24

## History

Originally observed by Messier in June 1764, M24 is not strictly speaking an open cluster. It is in fact a rich and vast concentration of part of the Milky Way, denser than the surrounding sky. Eighteenth-century astronomers, who did not have sufficiently accurate instruments, may well have thought that this region of the sky was a stellar object in its own right.

## Observation

This bright area of the Milky Way, next to the star γ Sagittarii and 6° northeast of the nebula M8, indicates the galactic center. M24 contains, in its northern part, 15′ from the star GC 24950, the open cluster NGC 6603. This stellar system, which is very dense and of low magnitude, can just be seen with a 115/900 reflector at ×72 magnification, and its members are not resolved. Observing M24 with a telescope is not of great value, apart from revealing the stellar richness of this region of the sky. M24, a vast field of blazing stars, is lost against the background sky; with

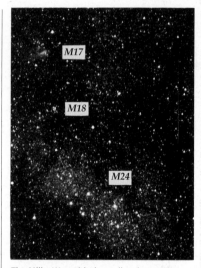

*The Milky Way with the stellar cluster M24 (see caption p. 43) (© Jean-Marc Lecleire).*

NGC 6603
Type d open cluster
NGC 6494
Epoch 2000.0 coord: α: 18 h 18 m 4 s
δ: −18° 25′

Apparent ∅: 5′
Absolute ∅: 20 l.y.
Mv: undetermined
Mp: 11.1
Distance: 16000 l.y.
Const.: Sagittarius
Favorable period: summer

OPEN CLUSTERS

M24. Milky Way Cloud
NGC 6494
Epoch 2000.0 coord: α: 18 h 16 m 5 s
                    δ: −18° 50′
Apparent ∅: 1° 35′×35′ − absolute ∅: 15 l.y.

Mv: 4.6 − Mp: 4.6
Distance: 16000 l.y.
Const.: Sagittarius
Discovery: Messier in 1764
Favorable period: summer

11×80 binoculars, however, the oberver is immersed in a sight of magical intensity.

*M24 Sgr, central region (field: see p. 34)*
(© J.-C. Merlin/Surelées Observatory).

# M25

## History

Chéseaux discovered M25 in 1746, 3° southeast of M24. The naked eye alone can locate this cluster in the nebulous veil of the Milky Way.

## Observation

M25 is very luminous in a 6×30 finder, and 15 or so stars can be resolved with 7×50 binoculars. A 115/900 reflector with a magnification of 100 (Or/HD 9-mm eyepiece) reveals much more: M25, bright and spread out, then fills the field of view, and some 50 stars can be resolved. The brightest of these, of almost eighth magnitude, are grouped in the center of the cluster. A 210-mm reflector, with a magnifi-

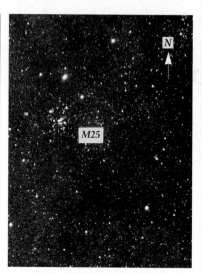

*The Milky Way with the stellar cluster M25 (see caption p. 43)* (© Jean-Marc Lecleire).

### CEPHEID STARS

These are intrinsic **variable stars** (see pp. 178 to 181), whose luminosity fluctuations are due to internal physical mechanisms. In 1908, the American astronomer Henrietta Leavitt detected almost 1800 **variables** of this type. In 1912, she found the period of 20 or so of these stars, and noticed that their brightness increased with their fluctuation period. From this, astronomers were able to map the Galaxy, by comparing the distance between two groups of stars, both of them containing at least one **Cepheid**. Shapley established the position of the Sun in the Galaxy in 1918, and Hubble identified M31 and M33 as extra-galactic objects.

| M25. Type d open cluster | Mv: 4.6 – Mp:4.6 – Distance: 2000 l.y. |
| IC 4725 | Const.: Sagittarius |
| Epoch 2000.0 coord: α: 18 h 31 m 6 s | Discovery: Chéseaux in 1746 |
| δ: −19° 15′ | Favorable period: June and July, in the middle |
| Apparent ∅: 32′ – absolute ∅: 20 l.y. | of the night |

N

*M25 Sgr (field: see p. 34)*
(© J.-C. Merlin/ Surelées Observatory).

cation of 50, does not enable any additional members of the cluster to be seen.

Extremely rarely for a galactic cluster, M25 contains one **Cepheid** star (δ Cephei type star): discovered by J. Schmith in 1866, U Sgr has a luminosity period of 6.74 days, with a magnitude varying between 6.4 and 7.1.

# M26

## History

Le Gentil discovered the galactic cluster M26 in 1750, and Messier catalogued it in 1764. This cluster is far from being as rich as its neighbor M11 and an average size instrument deceptively shows a rather poor stellar group.

## Search and observation

M26 is very easy to locate, 0.8° from the star δ Scutum, in an east–south-east direction, concealed in this very rich region of the Milky Way. It is 3.5° southwest of M11, and is clearly visible with 7×50 binoculars and appears as a small round cloudy patch whose diameter is close to one third that of the full Moon. A granular aspect difficult to make out against the very dense stellar background is revealed by 11×80 binoculars, while 20 or so stars up to magnitude 11 can be distinguished with a 115/900 reflector with a magnification of 150 (6-mm eyepiece). In all, M26 may contain fewer than 100 stars.

Despite its stellar poverty, M26 is very compact, and amateurs are advised to use higher magnifications than are usually used for open clusters. At a given magnification and 1.5 times the instrument diameter, the number of stars will not increase (M26 does not seem to go beyond twelfth magnitude), but more components will be resolved, and M26 will appear as a low density globular cluster with ill-defined borders.

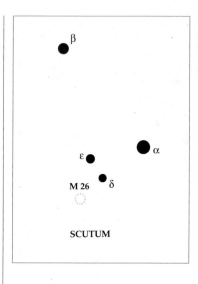

M26. Type f open cluster
NGC 6694
Epoch 2000.0 coord: α: 18 h 45 m 2 s
                    δ: −09° 24′
Apparent ⌀: 14′ – absolute ⌀: 12 to 16 l.y.

Mv: 8.0 – Mp:9.3
Distance: 4900 l.y.
Const.: Scutum (The Shield)
Discovery: Le Gentil in 1750
Favorable period: summer

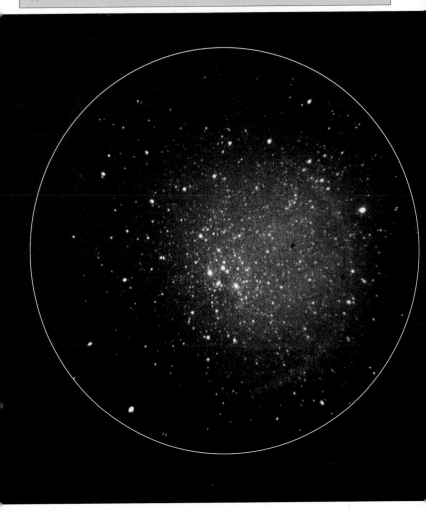

*M26 (field: see p. 34)* (© J.-C. Merlin/Surelées Observatory).

# M29

## History

Discovered by Messier in July 1764, M29 is located right in the middle of the Milky Way, 1.7°south–southeast of the star γ Cygni. This region of the sky is so dense and rich in bright stars that the novice observer might be discouraged from looking for this cluster. Nevertheless, amateurs will identify M29 easily with 7×50 binoculars or a 6×30 finder, when it looks like a nebulosity within which 5 or 6 stars are resolved.

*M29 is a perfect example of a poor and very open cluster. It is difficult to locate and its observation requires the use of very low magnification and very large field eyepieces (115/900 reflector, AH 400-mm eyepiece).* (Drawing: H. Burillier.)

## Observation

The cluster shows a poor and irregular structure in a 115/900 reflector with a 18-mm eyepiece (×50 magnification), and only 15 or so stars up to

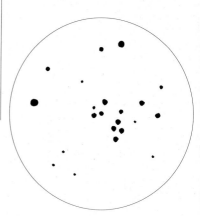

---

### CHOOSING AN EYEPIECE FOR THE OBSERVATION OF LOW CONTRAST OBJECTS WITH A 115/900 REFLECTOR

The observer can estimate the minimum magnification needed by dividing the instrument diameter by 5 (average diameter of the pupil at its maximum dilation). This coefficient varies of course with the age of the observer.

Minimum magnification: 115/5 = 23 times. The AH 40-mm eyepiece provides the longest focal length (40 mm) for these instruments, and its magnification is 900/40 = 22.5 times.

For an apparent field of 33° as given by the manufacturer, the absolute field equals the apparent field (33°) divided by the magnification (22.5), i.e. 1.46° or 1° 28′, i.e. three times the diameter of the full Moon!

M29. Type d open cluster – NGC 6913
Epoch 2000.0 coord: α: 20 h 23 m 9 s
δ: +38° 32′
Apparent ∅; 10′ – absolute ∅: 10 l.y.
Mv: 6.6 – Mp:7.1

Distance: 4000 l.y.
Age: 10 million years
Const.: Cygnus (The Swan)
Discovery: Messier in July 1764
Favorable period: summer

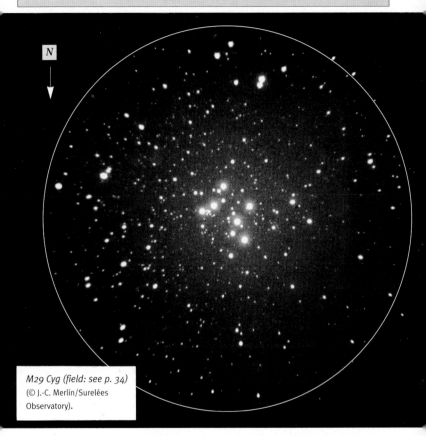

*M29 Cyg (field: see p. 34)*
(© J.-C. Merlin/Surelées
Observatory).

mv 9 or 10 are visible. Twenty stars will be resolved with a 200-mm reflector, but the observation is still not particularly spectacular, and defining the limits of the group becomes difficult. A Dobson-type reflector, or 11×80 binoculars will give the most interesting view of the cluster. M29 is a typical example of these poor and widely spread out objects for which the observer should concentrate on field of view and luminosity, using the lowest magnification of his or her instrument.

# M34

## History

Visible with the naked eye under very good seeing conditions, this galactic cluster was discovered by Messier in August 1764.

## Search

Locating M34 is fairly easy with binoculars, 5° west–northwest of β Persei (Algol), just at the border of the constellation Andromeda.

instruments are fitted with wide or ultra-wide field eyepieces (67° to 84°). Fifty stars are resolved with a 115/900 reflector, ×100 magnification, and 100 with a 20/1260 reflector, ×140 magnification. The center of M34 shows a beautiful concentration of small bright stars spread over 9°, with a population estimated at 80 members.

## Observation

M34 has a rather odd appearance, since its most luminous components (mv 7–7.5) form a cross. This stellar group, whose diameter is near 35′, is really enjoyable only with large binoculars, the ideal being 11×80 or 12×80. Owners of large aperture reflectors (F/d smaller or equal to 4.5) can also make nice observations if their

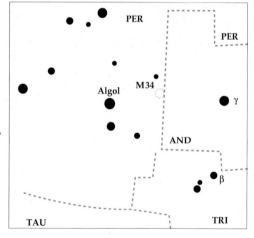

M34. Type d open cluster – NGC 1039
Epoch 2000.0 coord: α: 02 h 42 m 0 s
                          δ: +42° 47′
Apparent ∅: 35′ – absolute ∅: 18 l.y.
Mv: 5.2 – Mp: 5.5 – Distance: 1500 l.y.

Age: 100 million years
Const.: Perseus
Discovery: Messier in 1764
Favorable period: late October around midnight, near the zenith

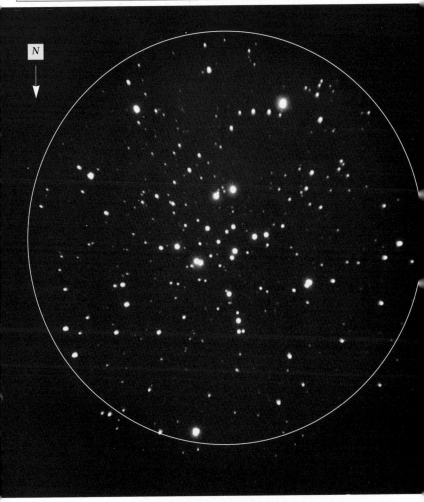

*M34 Per (field: see p. 34)*

(© J.-C. Merlin/Surelées Observatory).

# M35

## History

Observed for the first time by Messier in 1764, but first described by Chéseaux in 1746, M35 is a splendid open cluster located far west of the constellation Gemini. In the countryside, beyond any light pollution, it is easy to locate with the naked eye and looks like a circular patch of 30′ diameter.

## Observation

A few stars are already resolved in the finder. M35 contains over 300 blue, yellow and orange stars, of which the most brilliant are of

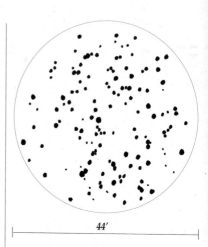

*44′*

*M35 with a 210-mm f:6 reflector, Or/HD 18-mm eyepiece, ×70 magnification, apparent and absolute fields of 52° and 44′ respectively* (Drawing: H.Burillier).

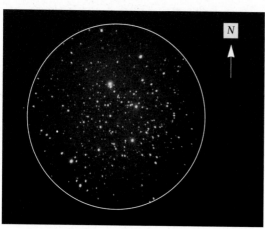

N

*M35 Gem (field: see p. 34)* (© J.-C. Merlin/Surelées Observatory).

M35. Type e open cluster
NGC 6913
Epoch 2000.0 coord: α: 06 h 08 m 9 s
δ: +24° 20′
Apparent ⌀: 30′ – absolute ⌀: 20 l.y.
Mv: 5.3 – Mp: 5.6 – Distance: 3000 l.y.

Age: 10 million years
Const.: Gemini
Discovery: Chéseaux in 1746
Favorable period: late December around midnight at the meridian passage

*M35 as seen with binoculars. Its circular form allows its unambiguous identification, and in a moonless night in the countryside, it can be located with the naked eye (photograph taken with a 135-mm telephoto-lens and an exposure time of 30 min).*
(© Jean-Marc Lecleire.)

magnitude 8.5. The image is striking with binoculars or a reflector with moderate magnification: M35 fills the whole field, and most stars are resolved. In this region of the sky close to the ecliptic, the Moon or a planet regularly passes the neighborhood of M35, and the amateur should not miss the chance to see or photograph these encounters.

# M36

## History

Discovered by Le Gentil in 1749, M36 is the first, and the brightest, of the three very bright galactic clusters in the constellation Auriga (The Charioteer). Very dense, it is very easy to locate with the finder or binoculars.

## Observation

The region close to the center is very compact, with an apparent diameter of 10', and is barely resolved with a small magnification 115/900 reflector. M36's borders are somewhat ill-defined, and the amateur ought to use relatively large magnifications (1 to 1.5 times the instrument diameter) and to observe the center of the cluster rather than its edges.

*View of M36 with a 115/900, ×50, reflector*
(Drawing: H.Burillier).

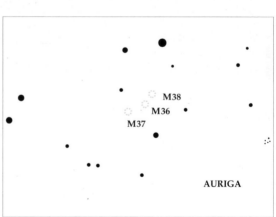

*Almost in alignment, M36, M37 and M38 are easy to locate with binoculars.*

M36. Type f open cluster
NGC 1960
Epoch 2000.0 coord: α: 05 h 36 m 2 s
δ: +34° 09'
Apparent ⌀: 10' – absolute ⌀: 14 l.y.

Mv: 6.3 – Mp: 6.5
Distance: 4100 l.y.
Const.: Auriga (The Charioteer)
Discovery: Le Gentil in 1749
Favorable period: autumn and winter

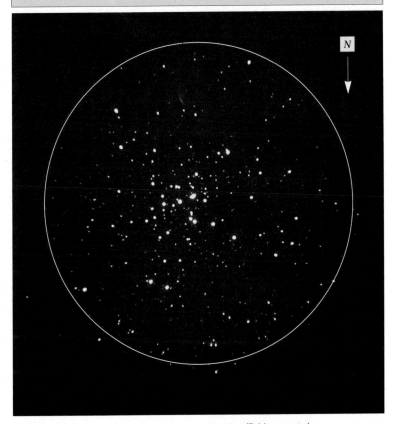

N

*M36 Aur (field: see p. 34)*
(© J.-C. Merlin/Surelées Observatory).

Fifty or so bluish stars are visible with 60-mm binoculars. M35 is a very young cluster rich in rapidly rotating **blue giants**. Its luminosity is 5000 or so times that of the Sun, which makes it comparable to the Pleiades.

# M37

## History

M37 was discovered by Messier on September 2, 1764. It is a splendid cluster, considered by some to be the most beautiful open cluster of the boreal sky. Its luminosity is about 2500 times that of the Sun.

## Observation

M37 is so dense and luminous that the novice can at first be mislead into identifying it as a globular cluster. M37 appears in the finder as a circular milky cloud, while a 115/900 reflector with an average magnification reveals a splendid stellar swarm of stars surrounded by a pale glow. Many additional stars of lower magnitude appear with a 210-mm reflector. M37 contains over 600 stars up to fifteenth magnitude, of which 150 are brighter than twelfth magnitude and can be observed with small diameter instruments.

*View of M37 with a 115/900 ×50 reflector. Although it is an open cluster, M37 is so dense that, even with a small diameter instrument it is difficult to resolve stars in the center.* (Drawing: H. Burillier.)

M37. Type f open cluster
NGC 2099
Epoch 2000.0 coord: α: 05 h 49 m 0 s
δ: +32° 33'
Apparent ∅: 25' – absolute ∅: 25 to 30 l.y.

Mv: 5.8 – Mp: 6.2 – Distance: 4600 l.y.
Age: about 200 million years
Const.: Auriga
Discovery: Messier in 1764
Favorable period: autumn and winter

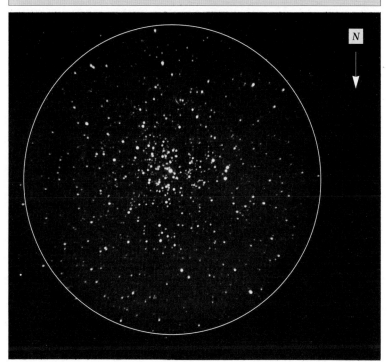

N

*M37 Aur (field: see p. 34)*
(© J.-C. Merlin/Surelées Observatory).

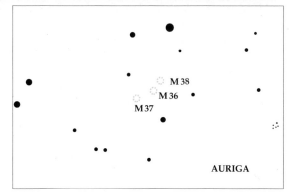

M 38
M 36
M 37

AURIGA

*Alignment of M36, M37 and M38.*

# M38

## History

Le Gentil discovered this galactic cluster, located 2.3° northwest of M36, in 1749. Although comparable in apparent and real size to M37, it is, however, much more populous.

## Observation

M38 appears as a diffuse nebulosity, clearly visible with a 6×30 finder, and its main components are resolved with a 115/900 ×50 reflector. Seventy or so visible stars are aligned along two lines in a north–south direction, which become even clearer with a 210-mm reflector. The cluster's edge then becomes lost in its very dense surroundings. A concentration of less luminous stars, of close to twelfth magnitude, is worthwhile observing at greater magnification (0.5 to 1 times the instrument diameter).

The faint cluster NGC 1907 lies 30′ southwest of M38, spreading over 6′ and visible with 11×80 binoculars in good seeing conditions. It is also visible with a 115/900 ×70 reflector, but cannot be resolved. Twelve or so stars can be resolved with a 210-mm, ×140, reflector, but the center remains diffuse.

*View of M38 with a 115/190 ×50 reflector.*
(Drawing: H. Burillier.)

NGC 1907
Type f open cluster
Epoch 2000.0 coord: α: 05 h 28 m 0 s
δ: +35° 17′

Apparent ⌀: 6′
Mv: 8.2
Mp: 9.9
Distance: 6800 l.y.
Const.: Auriga
Favorable period: autumn, winter

M38. Type e open cluster
NGC 1912
Epoch 2000.0 coord: α: 05 h 25 m 3 s
                    δ: +35° 48′
Apparent ⌀: 18′ – absolute ⌀ 21 l.y.

Mv: 6.8– Mp: 7.4 – Distance: 4200 l.y.
Age: 50 million years
Const.: Auriga
Discovery: Le Gentil in 1749
Favorable period: autumn, winter

*M38 Aur (field: see p. 34)* (© J.-C. Merlin/ Surelées Observatory).

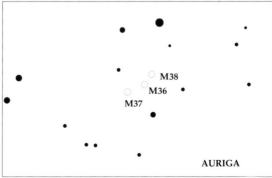

*Alignment of M36, M37 and M38.*

# M39

## History

Discovered by Le Gentil in 1750, the population of M39 is typical of that of open clusters. This splendid stellar group is found 9° northeast of the bright star Deneb (α Cygni), and can be easily spotted with the naked eye in good seeing conditions.

## Observation

The novice ought to start his or her search from Deneb, since finding this cluster amongst such a number of stars in the middle of the Milky Way can be tricky. Binoculars or a finder will allow the unambiguous identification of M39 as a triangular group made up of 20 or so stars. With 7×50 binoculars, this beautiful cluster, now isolated, is very spread out and brighter than most of its surrounding stars. Thirty or so members of magnitude 7 to 10 are resolved with a 115/900 reflector.

It is not necessary to use high magnifications for observing this splendid stellar formation, and priority should be given to field of view and luminosity.

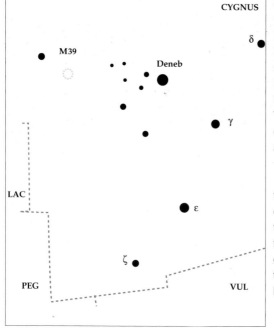

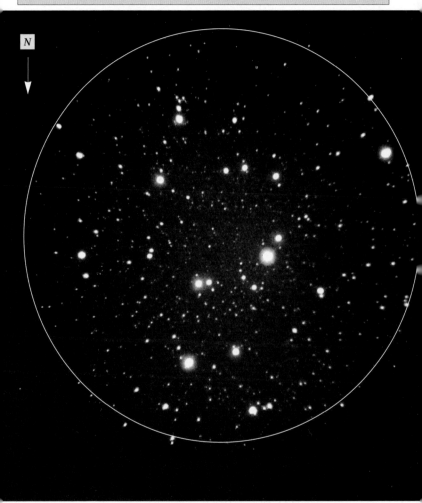

M39. Type e open cluster
NGC 7092
Epoch 2000.0 coord: α: 21 h 32 m 2 s
δ: +48° 26'
Apparent ∅: 32' – absolute ∅: 7 l.y.

Mv: 5.2 – Mp: 5.3
Distance: 800 l.y.
Const.: Cygnus
Discovery: Le Gentil in 1750
Favorable period: summer

N

*M39 Cyg (field: see p. 34)*
(© J.-C. Merlin/Surelées Observatory).

# M41

## History

M41 is a splendid galactic cluster discovered by Flamsteed in 1702, and catalogued by Messier in 1765. Aristotle is said to have seen this faint diffuse patch in 325 BC.

## Search

This beautiful cluster, 4° south of the white and luminous star Sirius, contains 25 bright stars, up to sev- enth magnitude, and has an appar- ent diameter the size of the Moon. For this reason M41 is a favorite with beginners.

## Observation

Visible with the naked eye, the stel- lar group is resolved with 11×80 binoculars. Near the center, a sev- enth magnitude red–orange star makes a nice contrast with its fainter neighbors. Binoculars of 60 mm with a magnifi- cation of 56 show 40 to 50 resolved objects, and a 115/900, ×72, reflector resolves more. Sparse and irregular, M41 spreads over more than 30′ and entirely fills the field of view; 150 members can be

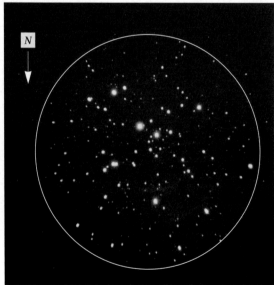

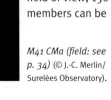

*M41 CMa (field: see p. 34)* (© J.-C. Merlin/ Surelées Observatory).

*M41* (© J.-F. Leoni/*Ciel et Espace*).

isolated, of magnitudes 8 to 11.5.
The astronomer will see a magical
image with a 355-mm (Celestron 14)
Schmidt–Cassegrain reflector,
revealing over 250 stars of up to
fourteenth magnitude!

# M44

Other name **PRAESEPE**

M44 is a splendid and famous open cluster in Cancer. Known for ages, it owes its importance to its membership of the **zodiac**.

## Observation

Of apparent visual magnitude 3.7, this group can only be seen with the naked eye under good seeing conditions. M44 is a very rich cluster, and contains at least 2300 stars; unfortunately, most of them are of magnitudes greater than 15 and can only be seen with large diameter instruments. Thirty or so stars can be resolved with binoculars. The lowest magnifications, covering the largest fields, give the most favorable observations. The most impressive is obtained with 11×80 or 12×80 binoculars, although 7×50 binoculars also lead to very good results.

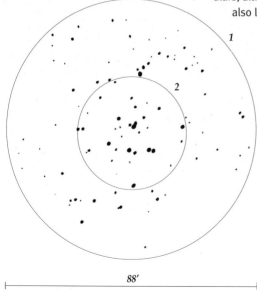

88′

*M44 with a 115/900 reflector. Circle 1: AH 40-mm eyepiece, with a magnification of 22.5 and a field of 88′; circle 2: Or/HD 9-mm eyepiece, ×100, field of 31′.*
(Drawing: H.Burillier.)

| M44 – Praesepe | Apparent ∅: 1°30′ – absolute ∅: 13 to 40 l.y. |
| --- | --- |
| Type d open cluster | Mv: 3.7 – Mp: 3.7 |
| NGC 2632 | Distance: 525 l.y. |
| Epoch 2000.0 coord: α: 08 h 40 m 1 s | Const.: Cancer (The Crab) |
| δ: +19° 59′ | Favorable period: mid January, at midnight |

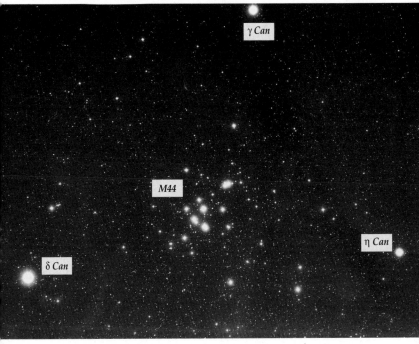

*M44 Praesepe* (© J.-M. Lopez/A.M. Jacquey/
Pises Observatory).

An AH 40-mm eyepiece is ideal for a 60-mm refractor or a 115/900 reflector, and gives a wide field of view of low magnification, which is both very luminous and with strong contrast. For instruments of diameter of 200 mm or more, wide and ultra-wide field eyepieces enhance the splendor of the image as well as its resolution.

# M45

Other name **THE PLEIADES**

## History

An outstanding and splendid open cluster located in Taurus (The Bull), 12° northwest of the Hyades group, the Pleiades have been known and observed from time immemorial. In his calendar, Caesar fixed the beginning of summer at their **heliacal** rise; Homer mentioned them in the *Odyssey*, while the Aztecs invested them with magical power.

## Observation

A trained naked eye can distinguish 5 or 6 stars. Mästlin, Kepler's private tutor, is said to have counted a dozen. Observing the Pleiades with binoculars offers a spectacle of rare

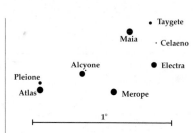

beauty. A hundred or so stars are then visible. The predominantly bluish or whitish tints attest that all the stars were born in the same period, from the heart of the same primary nebula. Attentive observation of the region south of Merope with a reflector of at least 200-mm aperture will allow the not-yet-dissipated veils of the primary nebula to be seen.

The Pleiades cluster is indeed very young: it is estimated to be a mere 40 million years old. Thermonuclear reactions

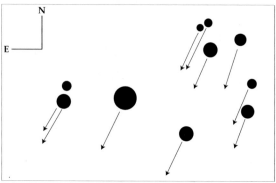

*The Pleiades' proper motion over a period of approximately 20 000 years.*

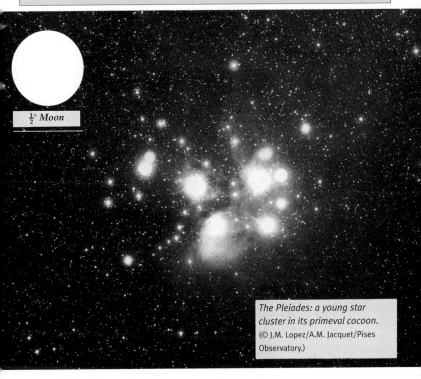

The Pleiades
Type c open cluster
Messier number: M45
Epoch 2000.0 coord: α: 03 h 43 m 9 s
δ: +23° 58′

Apparent ⌀: 2° – absolute ⌀: 20 l.y.
Distance: 400 l.y.
Age: 40 million years
Const: Taurus
Favorable period: winter

½° *Moon*

*The Pleiades: a young star cluster in its primeval cocoon.*
(© J.M. Lopez/A.M. Jacquet/Pises Observatory.)

have just started in the core of the stars it is made of. Most of these stars are rapidly rotating, showing that they have not yet reached their maturity. In particular, Pleione, rotating rapidly at 200 km/s, ejects from time to time a large amount of gas. Analysis of its light-curve suggests that its atmosphere is still subject to motion due to intense turbulence, and the drop in luminosity can reach half a magnitude. The most recent phenomenon of this kind was studied in 1972. M45 should preferably be observed with binoculars, or any other wide-field instrument.

Alcyone is the most brilliant body of the group (magnitude of 2.86) before Atlas (3.67), Electra (3.71), Maia (3.88), Merope (4.18) and Taygete (4.31).

# M46

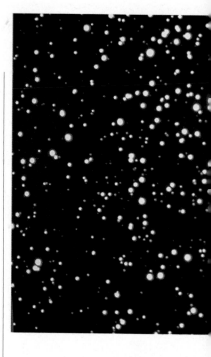

## History

Charles Messier discovered this galactic cluster on the 19th of February 1771.

## Search

M46 is 14° 30' east of the bright star Sirius, and is clearly visible with 7×50 binoculars. It appears as a circular, dense and diffuse cluster.

## Observation

Sixty or so stars are resolved with a 115/900 reflector and a 9-mm eyepiece, whose magnification (100 times) allows the observer to separate many stars close to each other near the cluster's center. M46 fills the whole field of view, and a 153-mm refractor will resolve a hundred or so stars.

M46 (NGC 2437) contains the small planetary nebula NGC 2438; this celestial curiosity, whose apparent diameter (only 68″) is slightly larger than that of Jupiter, is particularly difficult to locate in instruments of diameter smaller than 200 mm. Lost in the brightness of the neighboring stars, at M46's edge and 7' from its center, NGC 2438 was discovered by William Herschel at the beginning of the nineteenth century. The observer is strongly advised to use an oxygen III (OIII) filter, which will enhance the contrast between this 11th magnitude planetary nebula and the background sky, while stars of low and average magnitude will appear fainter.

M46 – Type f open cluster
NGC 2437
Epoch 2000.0 Coord: α: 07 h 41 m 8 s
δ: −14° 49′
Apparent ⌀: 27′ – absolute ⌀: 30 l.y.
Mv: 6.1 – Mp: 6.6

Distance: 4700 to 5400 l.y.
Const.: Puppis (The Poop or Stern)
Discovery: Messier in 1771
Favorable period: December and January, in
the middle of the night

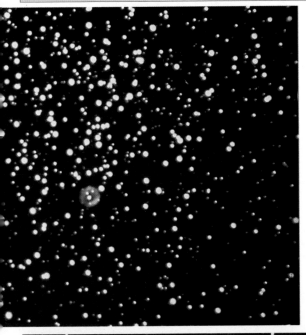

The planetary
nebula NGC 2438
is located at the
edge of the galactic
cluster M46.
(© J.S. Devaux/*Ciel et
Espace*.)

M46 and NGC 2438
Puppis: photograph
taken on the 25th of
December 1992 with
the 1 m reflector
at Puimichel
(© J.-M. Lecleire).

# M47

## History

Messier discovered this splendid galactic cluster, only 1° 30′ west of M46, in February 1771. Clearly visible with the naked eye on a very dark night, M47 is one of the most spread out and irregular open clusters. Most of its stars are very bright (magnitudes 6, 7 and 8) and are widely dispersed over an area comparable to that of the full Moon (a diameter of 0.5°).

## Observation

Using a 115/900 reflector and a magnification of 50 (Or/HD 18-mm eyepiece), M47 fills the field of view (absolute field: 62″) and looks very open. It is difficult to find the limits of the group using a 210-mm reflector and a magnification of 80, where over 60 stars are visible. M47 contains two splendid binary stars that binoculars of at least 60-mm diameter may separate: Σ1120 (components of mv 6.0 and 9.5, separated by 17.5″) and Σ1121, a close binary (components of mv 7 and 7.5, separated by 7.4″).

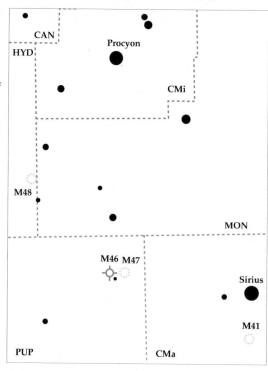

M47. Type d open cluster
NGC 2422
Epoch 2000.0 coord: α: 07 h 36 m 6 s
δ: −14° 30'
Apparent ⌀: 30' – absolute ⌀: 17 l.y.
Mv: 4.4 – Mp: 4.5 – Distance: 1700 l.y.

Age: 20 million years
Const.: Puppis (The Poop or Stern)
Discovery: Messier in 1771
Favorable period: December in the middle of the night

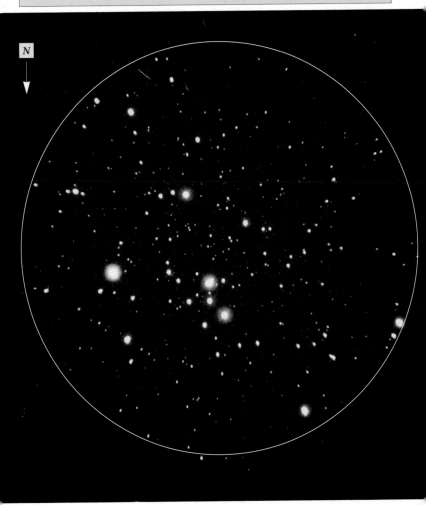

N

*M47 Pup (field: see p. 34)*
(© J.-C. Merlin/Surelées Observatory).

73

# M48

## Search

M48 is sufficiently spread out to be located with the naked eye, and appears like a blurred star of average luminosity using a 6×30 finder or 7×50 binoculars. The observer will easily find it from the noticeable equilateral triangle it forms with the stars C Hydri and ζ Monocerostis.

## Observation

The grainy aspect of this stellar formation is revealed by 11×80 binoculars, while binoculars of larger diameter provide without doubt the most interesting image for the observer, as the group is so widely spread.

As long as the amateur does not choose magnifications larger than 30 times, the 115/900 reflector

also provides good results, and in addition the observer is advised to use AH 40-mm and Or/HD 25-mm eyepieces. Beyond this, the magnification is too large and the cluster becomes almost indistinguishable against the bright background sky.

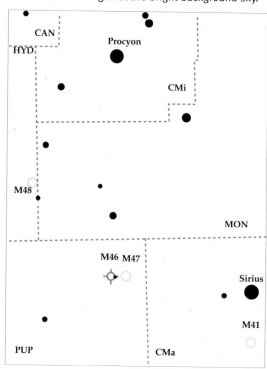

M48. Type f open cluster
NGC 2548
Epoch 2000.0 coord: α: 08 h 13 m 8 s
δ: −05° 48′
Apparent ⌀: 54′ – absolute ⌀: 20 l.y.

Mv: 5.8 – Mp: 5.3
Distance: 1700 l.y.
Const.: Hydra (The Sea Monster)
Discovery: Messier in 1771
Favorable period: winter

*M48 Hya (field: see p. 34)*

# M50

## History

J.-D. Cassini first reported this group of stars in 1711, and Messier described it as a "more or less bright" cluster in April 1772.

## Search

The amateur will easily locate this beautiful galactic cluster along the line between Sirius and Procyon. Using 7×50 binoculars, the stellar formation appears circular, and at the cluster's edge a few stars will be resolved.

*Locating M50 from Rigel (α Ori) on the same declination circle.*

## Observation

With a 115/900 reflector and a magnification of 72 (12.5-mm focal length eyepiece), a formation of stars in an arc appears at the cluster's edge; sixty or so stars are visible, and some 15 stars of up to tenth magnitude are resolved.

M50 presents a very rich field covering 16′ where over 100 glittering points of light reveal themselves in a 153-mm refractor, and using a 355-mm Schmidt–Cassegrain (C14) reflector and a magnification of 400 over 200 shine. The luminosity of the globular cluster M50 is estimated to be 1600 times that of the Sun.

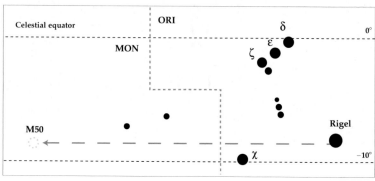

M50. Type e open cluster
NGC 2323
Epoch 2000.0 coord: α: 07 h 03 m 2 s
                    δ: −08° 20′
Apparent ⌀: 16′ − absolute ⌀: 10 l.y.

Mv: 5.9 − Mp: 6.9 − Distance: 2900 l.y.
Const.: Monoceros (The Unicorn)
Discovery: J.-D. Cassini in 1711
Favorable period: winter in the first half of the night

*M50 (field: see p. 34)* (© J.-C. Merlin/Surelées Observatory).

# M52

## History

Messier is thought to have discovered this stellar group on the 7th of September 1774, while observing Comet Montaigne.

## Search

M52 is a splendid galactic cluster located at the edge of the constellation Cepheus. The novice observer will locate it without difficulty using binoculars, by extending the segment made up of α and β Cassiopeiae by just over twice the distance between these two stars. The astronomer will then see a small nebulosity, slightly elongated and very compact.

## Observation

A hundred or so stars of magnitude 9 to 10 can be observed with a 115/900 reflector and a magnification of 130 (7-mm eyepiece). The region near the cluster's center is still very dense, and M52 reveals its granular aspect against a whitish, rather misty, background. A 210-mm reflector provides the astronomer with an entrancing picture, with just under 200 stars.

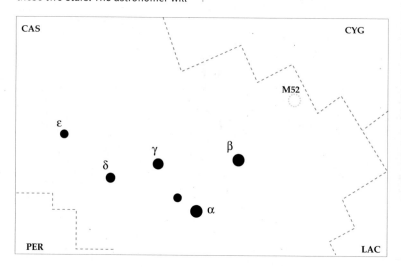

M52. Type e open cluster
NGC 7654
Epoch 2000.0 coord: α: 23 h 24 m 2 s
δ: +61° 35'
Apparent ∅: 12' – absolute ∅: 20 l.y.
Mv: 6.9 – Mp: 7.3

Distance: 5000 to 8000 l.y.
Age: 50 million years
Const.: Cassiopeia
Discovery: Messier in 1774
Favorable period: autumn in the middle of
the night (passage at the superior meridian)

N

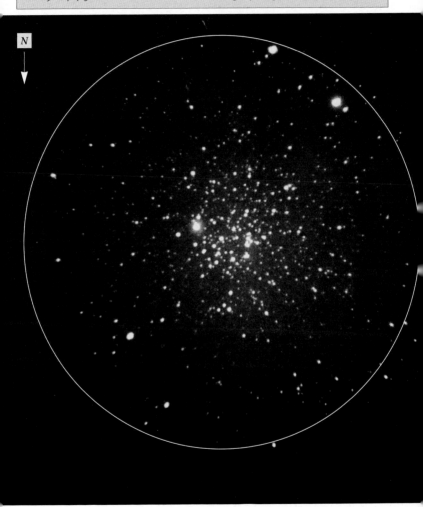

*M52 Cas (field: see p. 34)*
(© J.-C. Merlin/Surelées Observatory).

# GLOBULAR CLUSTERS

# M2

## History

Apart from the famous Helix Nebula (see p. 126), the constellation Aquarius has an exceptional globular cluster discovered by Giovanni Domenico Maraldi in 1746 and catalogued by Messier a few years later.

## Search

Next to the celestial equator, M2 is so bright and dense that observing it does not require any specialized equipment. The astronomer will easily locate it from the star β Aquarii, on the same hour circle. In practice, the right ascension axis of an equatorial instrument is fixed, and only the declination motion is used. The observer then simply moves the instrument 5° northward and finds M2. With a 6×30 finder or 7×50 binoculars, M2 looks like a blurred sixth magnitude star.

## Observation

A 115/900 reflector with a magnification of 100 reveals all the splendor of M2, whose granular aspect, typical of this type of cluster, is revealed against a vast luminous peripheral halo of stars, of which over 30 may be resolved. Using a 210-mm reflec-

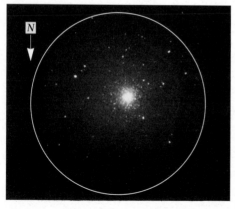

M2 Aqr. (© J.-C. Merlin/Surelées Observatory). *The field of this photograph is 20"/mm, i.e. 30' (40-cm reflector with an image intensifier) and represents the field obtained using a 115/900 reflector with a 9-mm Or/HD eyepiece.*

| | |
|---|---|
| M2. Type II globular cluster | Distance: 40 000 l.y. |
| NGC 7089 | Const.: Aquarius |
| Epoch 2000.0 coord: α: 21 h 35 m 5 s | Discovery: Maraldi in 1746 |
| δ: −00° 49′ | Favorable period: late summer in the second |
| Apparent ⌀: 12′ – absolute ⌀: 120 l.y. | half of the night |
| Mv: 6.4 – Mp: 5 | |

M2

*M2 with a CCD camera*
(© C. Buil/*Ciel et Espace*).

tor and a magnification of 130, the observer will be able to see peripheral stars up to thirteenth magnitude, as well as faint zones nearer the cluster's center. M2 is entirely resolved with a 355-mm Schmidt–Cassegrain reflector. The spherical aspect of this object may be seen with anything from the smallest of instruments, such as a 60-mm refractor. In all cases the image remains mostly whitish in color, although under very good seeing conditions the amateur sometimes sees slightly bluish tints.

Lovers of **variable stars** will enjoy studying these interesting objects, which are present in large numbers in globular clusters. Among others they will find **Cepheid**, **RR Lyrae** and **RV Tau** type stars, which are not consistently studied by amateur observers of variable stars because of their low amplitude variation and very short periods.

# M3

## History

Next to the north galactic pole, M3 is one of the most beautiful globular clusters of the boreal sky. It was discovered by Messier in 1764.

## Search

At a latitude roughly half way up the northern hemisphere, M3 is high enough in the sky, approaching the zenith where seeing conditions are optimal. At the limit of the constellations Boötes and Coma Berenices, it is easy to locate with 7×50 binoculars; the observer has simply to imagine an isosceles triangle pointing northwest, with its base the stars ε and β Boötis.

## Observation

Trained naked eyes will be able to see M3 under a moonless sky at high altitudes. Easily recognizable using 7×50 binoculars, the cluster reveals its grainy nature with 11×80 binoculars, without, however, being resolved. A 115-mm reflector with an magnification of 100 is necessary to partially isolate stars up to twelfth magnitude at the edges. Using a 200-mm reflector and a magnification of 140, an ultra-wide field eyepiece of 84° of apparent field reveals an image of rare beauty: M3 suddenly looks like a myriad of Suns which fills the whole field of view. The center is extremely dense: several hundred stars are visible. Photographs obtained at the 5-m giant telescope at Mount Palomar in the 1950s showed that M3 has a diameter extending to 20′. Up to 45 000 stars, of a brightness equal to or greater than photographic

*M3 CVn (field: see p. 82)*
(© J.-C. Merlin/Surelées Observatory).

M3. Type VI globular cluster
NGC 5272
Epoch 2000.0 coord: α: 13 h 42m 2 s
                    δ: +28° 23′
Apparent ⌀: 10′ – absolute ⌀: 220 l.y.
Mv: 6.4 – Mp: 4.5

Distance: 35 000 to 40 000 l.y.
Const.: Canes Venatici (The Hunting Dogs)
Discovery: Messier in 1764
Favorable period: April in the middle of the night

*M3* (© AFA/*Ciel et Espace*).

magnitude 22.5, were counted up to 8′ from the center.

M3 might contain half a million stars, among which are a very large number of **variable stars**, notably **RR Lyrae**, whose fluctuation periods vary between 10 min and 12 h.

# M4

## History

Discovered by Chéseaux in 1746 and catalogued by Messier in 1764, the globular cluster M4 is rather difficult to observe at latitudes half way up the northern hemisphere.

## Search

Always very low on the horizon, this object is therefore very affected by atmospheric turbulence. However, the observer can locate it using 7×50 binoculars, in the reddish glow of the bright star Antares, which is located only 1° 18′ to the east.

## Observation

With a 115/900 reflector equipped with a 12.5-mm focal length eyepiece, giving a magnification of 72, M4 appears as a small and rather faint patch. It does not have the typical spherical aspect of globular clusters, and seems to dissolve rather rapidly

into the surrounding space. Because of this, the novice observer unfamiliar with these objects might think that it belongs to the category of open clusters. However, to astronomers observing M4 from the austral hemisphere when it is near the zenith, it would be comparable in its density and brightness to M13 in the Hercules.

A large number of stars up to twelfth magnitude are resolved using the same 115-mm reflector and a magnification of 128. With such a great magnification, the brightness due to the relatively close star

*M4 Sco (field: see p. 82)* (© J.-C. Merlin/Surelées Observatory).

M4. Type IX globular cluster
NGC 6121
Epoch 2000.0 coord: α: 16 h 23 m 7 s
            δ: −26° 31′
Apparent ⌀: 22′ – absolute ⌀: 50 l.y.

Mv: 6.5 – Mp: 6.8
Distance: 5700 l.y.
Const.: Scorpius (The Scorpion)
Discovery: Chéseaux in 1746
Favorable period: July early in the night

*M4 or NGC 6121*
(© C. Buil/*Ciel et Espace*).

Antares is reduced, and the stellar group then shows an extraordinary stellar richness, with a line of stars cutting the cluster's center into two distinct lobes. A 210-mm reflector will not reveal more.

With NGC 6397 in the austral constellation Ara, M4 is our nearest galactic globular cluster, in which a large number of **RR Lyrae** type **variable stars** have been detected.

# M5

## History

G. Kirch discovered this globular cluster in 1705. M5 is one of most extraordinary stellar concentrations in the sky, and in the last century Camille Flammarion wrote of this celestial beauty that "its richness of stars is so great that counting them must be impossible".

## Search

In very good seeing conditions, the amateur astronomer can see M5 with the naked eye; the cluster is located in the Serpent's head, 20′ north–northwest of the double star 5 Serpentis.

## Observation

Using a 6×30 finder or 7×50 binoculars, the observer recognizes M5's star-studded appearance, typical of globular clusters. With 11×80 binoculars M5 is revealed as one of the most radiant clusters of the boreal sky, with M13 and M3.

A few stars are easily resolved towards the edges of the cluster. Using a 115/900 reflector and medium to high magnifications, M5 looks rather like M13: the dense and extremely luminous core looks like a bar of white stars. A careful study shows that M5 is not perfectly spherical, but is flattened by some 10%, which could imply a rotation of the whole formation.

The astronomer can observe over 100 stars with a 210-mm reflector and a magnification of 200. The image is very striking with large aperture instruments: with an aperture of at least 250 mm, over 300 stars can be counted, although they still appear very close to each other, and 31.75-mm diameter ultra-wide field eyepieces reveal a magical image. I will mention here the exceptional UWA 14-mm eyepiece (from Maede) and/or the 12-mm Nagler II eyepiece (from Télévue). For observers using 24.5-mm eyepieces, magnificent observations are possible with the Or/HD 12.5-, 9- and 7-mm eyepieces (from Stellarion and/or Vixen).

*M5* (© T. Gregory/CFHT/*Ciel et Espace*).

M5. Type V globular cluster
NGC 5904
Epoch 2000.0 coord: α: 15 h 18 m 6 s
                    δ: +02° 05′
Apparent ⌀: 17.4′ absolute ⌀: 220 l.y.
Mv: 5.7 – Mp: 6.6

Distance: 27 000 l.y.
Const.: Serpens (The Serpent)
Discovery: Kirch in 1705
Favorable period: April–May in the second
half of the night

# M9

## Search

M9 is a cluster luminous enough to be recognized with a 6×30 finder. With 7×50 binoculars, its location can be found by a systematic search slightly east of the segment made by the two stars θ and η Ophiuchi, at a third of the distance from the star η Sabik.

## Observation

Observation with 11×80 binoculars reveals a circular structure of an extreme and even density, from the cluster core to the outer regions.

M9 cannot be resolved with a 115mm reflector, since its most brilliant stars are of magnitude 13.1. Using instruments of at least 200-mm aperture with a relatively high magnification (1.5 times the instrument diameter), 20 or so stars can be resolved at the cluster edge. The core is no longer smooth: the observation now shows it as very compact and grainy.

M9 shines as brightly as 60 000 Suns! Here again, a great number of **variable stars** have been detected. Using low magnification eyepieces, the observer can note, southwest of M9, the presence of strongly contrasting dark regions in this very rich part of the celestial sphere.

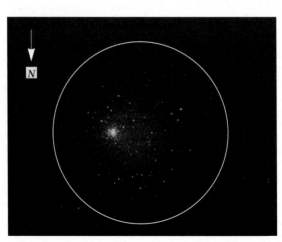

*Left: M9 Oph (field: see p. 82)* (© J.-C. Merlin/ Surelées Observatory).

*Right: CCD image of M9* (© C.Buil/*Ciel et Espace*).

M9: Type VIII globular cluster
NGC 6333
Epoch 2000.0 coord: α: 17 h 19 m 1 s
δ: −18° 31'
Apparent ⌀: 2.4' – absolute ⌀: 60 l.y.

Mv: 8 – Mp: 8.7
Distance: 6000 l.y.
Discovery: Messier in May 1764
Favorable period: summer

# M10

## Search

This magnificent perfectly circular cluster of suns is easily recognizable from the stars ε and ζ Ophiuchi, with which M10 forms an isosceles triangle.

## Observation

M10 is less dense than its companion M9 but is nevertheless interesting to study even with the smallest instruments. A 60-mm refractor with a magnification of 60 will already reveal the structure of the cluster's core, which begins to be resolved using a 115-mm reflector with a magnification of 100. The attentive observer will note that this stellar group, which extends slightly over 12′, is very spread out towards the edges, and that its surroundings are studded with a few isolated stars which probably do not belong to the cluster itself.

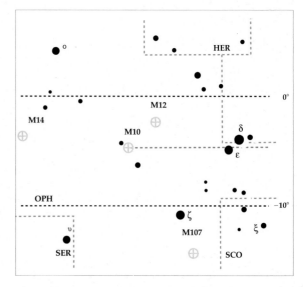

*M10 belongs to the same declination circle as the star δ Oph, which facilitates its search.*

M10. Type VII globular cluster
NGC 6254
Epoch 2000.0 coord: α: 16 h 57 m 1 s
δ: −04° 07′
Apparent ∅: 8.2′ − absolute ∅: 200 l.y.

Mv: 6.7 − Mp: 7.6
Distance: 16 000 l.y.
Const.: Ophiuchus (The Serpent Bearer)
Discovery: Messier in 1764
Favorable period: summer

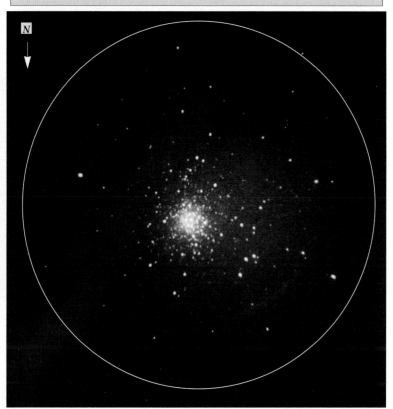

*M10 Oph (field: see p. 82)*
(© J.-C. Merlin/Surelées Observatory).

M10 may be entirely resolved with a 210-mm reflector; as soon as wide field eyepieces are used, it is seen in all its striking diamond-white luminosity. The astronomer then notes a number of dark areas towards the northern and southern regions of the formation. M10 seems to have fewer **variable stars** than most other globular clusters.

# M12

## Search

Using a 6×30 finder, the amateur easily recognizes M12 as a very luminous circular patch, which is very similar to M10, and which lies 3.4° to the southeast. It is preferable, however, to locate this nice cluster, using binoculars, on the segment made by β Kelb Alrai and ε Yed Posterior Ophiuchi, at a third of the distance from ε.

## Observation

Observations with a 115-mm reflector and a 12.5-mm Or/HD eyepiece (with a magnification of 72) reveal it as slightly ovoid. At sufficiently great magnifications, several stars are separable from the group, against a particularly poor background sky. Thirty or so stars on the periphery may be isolated using a 200/2000 reflector and a magnification of 140; the core reveals its grainy nature but remains very diffuse, while its luminosity contrasts strongly with the grayish external zones. The brightest stars are of eleventh magnitude, and one star

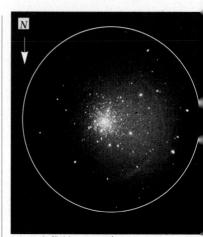

*M12 Oph (field: see p. 82)*
(© J.-C. Merlin/Surelées Observatory).

south of the cluster is brighter than all the others.

M12 contains ten or so **variable stars**. Its distance from our solar system is still controversial, but it seems that it may be just over 16 000 light-years away. If this is so, M12 would then be separated from M10 by 2000 light-years or so.

M12. Type IX globular cluster
NGC 6218
Epoch 2000.0 coord: α: 16 h 47 m 2 s
δ: −01° 57′
Apparent ⌀: 9.3′

Mv: 6.7 – Mp: 7.5
Distance: 16 000 to 24 000 l.y. ?
Const.: Ophiuchus (The Serpent Bearer)
Discovery: Messier in May 1764
Favorable period: summer

*CCD image of M12* (© C. Buil/*Ciel et Espace*).

# M13

## History

Amongst globular clusters visible at latitudes half way up the northern hemisphere, M13, in the constellation Hercules, is without any doubt the most frequently observed. But, strangely enough, no-one mentioned this cluster before Halley, who in 1714 catalogued it as a nebula just visible with the naked eye. Half a century later, Messier listed this object in his catalog under the heading "round nebula without any star".

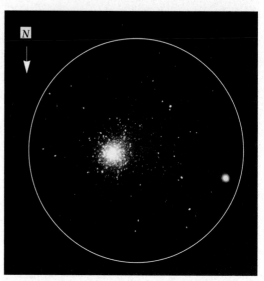

*M13 Her (field: see p. 82)* (© J.-C. Merlin/Surelées Observatory).

## Search

Located in the segment made by η and ζ Herculis, M13 is considered to be one of the most impressive clusters, along with M3 and M22. Its pale glow is visible with the naked eye in the countryside on a clear moonless night.

## Observation

With a 115-mm reflector, M13 exhibits a granular structure at its periphery, where over 20 clearly resolved stars encircle the diffuse and perfectly spherical central region. Instruments of average diameter resolve this part of the cluster. The observer is then advised to increase the magnification by a

M13. Type V globular cluster
NGC 6205
Epoch 2000.0 coord: $\alpha$: 16 h 41 m 7 s
                    $\delta$: +36° 27′
Apparent $\varnothing$: 10′ – absolute $\varnothing$: 175 l.y.

Mv: 5.7 – Mp: 6.4
Distance: 20 000 to 25 000 l.y.
Const.: Hercules
Discovery: Halley in 1714
Favorable period: summer

*M13* (© USNO/*Ciel et Espace*).

further 1.5–2 times the diameter, sufficiently reducing the luminosity of the cluster's center so that it appears as a sphere made up of minute dots of light.

Using a 210-mm reflector and a magnification of 140, M13 becomes a white haze of stars reminding one of a spider spreading four of its legs into the surroundings. A 530-mm or 355-mm reflector (C14) shows an unforgettable image, where M13, so bright and spread out, fills the whole field. This is certainly one of the most impressive sights of the night sky, a cluster of extraordinary density and richness which may contain almost half a million suns!

# M15

## Search

M15 is located 4° northwest of the star ε Pegasi, at the edge of Equulei (The Foal). It is one of the most interesting clusters in this sparse region of the sky. This beautiful concentration of stars is recognizable using a 6×30 finder as well as with 7×50 binoculars. However, it is one of the most difficult to isolate and requires instruments of average or large diameter to be observed properly.

## Observation

With a 115/900 reflector and a magnification of 100, M15 makes a good comparison with M13. It has a perfectly circular form, with a tight (2′ diameter) and very bright core. The observer will see 15 or so resolved stars towards its edge. A 200-mm reflector with a 14-mm UWA eyepiece, revealing a few additional stars, will give an entrancing image: many white

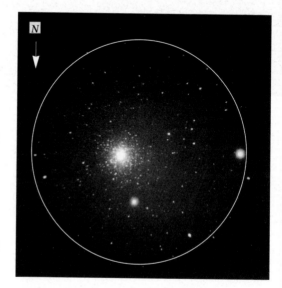

*M15 Peg (field: see p. 82)*
(© J.-C. Merlin/Surelées Observatory).

M15. Type IV globular cluster
NGC 7078
Epoch 2000.0 coord: α: 21 h 30 m 0 s
          δ: +12° 10′
Apparent ⌀: 7.4′ – absolute ⌀: 130 l.y.
Mv: 6.5– Mp: 7

Distance: 30 000 to 40 000 l.y.
Const.: Pegasus (The Winged Horse)
Discovery: Maraldi in 1745
Favorable period: summer in the middle of
the night

*M15 is so dense and so rich that the observer should use instruments of long focal length to study it properly. High magnification binoculars are preferable to reflectors, which sometimes have too short a focal length, precluding the use of high magnifications.* (© CFH/*Ciel et Espace*.)

and bright stars are resolved on the edge of the core, which is still seen as dense and diffuse on a milky background.

Observing M15 requires a magnification high enough, whatever the instrument; eyepieces giving magnifications up to 1.5 times the diameter lead to the best results.

As with most globular clusters, M15 contains a large number of variable stars, and over a hundred, mostly of **RR Lyrae** type, have been detected.

**99**

# M22

## History

This splendid cluster was observed for the first time in the middle of the seventeenth century, and catalogued by Messier in 1764. Only ω Centauri and 47 Tucanae outdo this exceptional cluster in luminosity and area.

## Search

M22 is near the ecliptic, 2° 3′ northwest of λ Sagittarii. This beautiful cluster is often considered by astronomers to be one of the grandest globular clusters visible at latitudes half way up the northern hemisphere.

## Observation

Visible with the naked eye under good seeing conditions, M22 can be located using a 6×30 finder, and 11×80 binoculars resolve a few stars on a grainy cream-colored background. The core of M22 is very dense, and provided the binoculars are stabilized by a tripod, the observer can see almost 30 stars at its periphery.

A 115/900 reflector with a 9-mm Or/HD eyepiece reveals a multitude of white stars, but M22 is, however, so rich and dense that the central

*M22 Sgr (field: see p. 82)* (© J.-C. Merlin/ Surelées Observatory).

M22. Type VII globular cluster
NGC 6656
Epoch 2000.0 coord: α: 18 h 36 m 4 s
δ: −23° 54′
Apparent ∅: 17.3′ – absolute ∅: 50 l.y.

Mv: 5.1– Mp: 6.3 – Distance: 22 000 l.y.
Const.: Sagittarius
Discovery: Ihle in 1665
Favorable period: July in the first half the
night

*M22, one of the richest globular clusters
of the Galaxy. The dominant blue and
white colors of this splendid cluster are
unmistakable. This photograph was taken
with a 203-mm Schmidt–Cassegrain
reflector, f:6.3 and an exposure time of
only 10 minutes.* (© Riffle/*Ciel et Espace*.)

regions remain unresolved. The
image is indescribably entrancing
with a 210-mm reflector and a magnification of 140: the cluster is
almost entirely resolved, even more
so than M13 in the constellation
Hercules. The eye wanders through
a myriad of dancing points of light,
and M22 is one of these celestial
splendors that the astronomer will
never tire of.

Like its companions, M22 contains
many **RR Lyrae** type **variable stars**,
and one **Mira** type star (see p. 176),
which varies between magnitude 14.5
and 17.5 with a period of 200 days.

# M28

## Search

M28 is located 45′ northwest of the star λ Sagittarii. It is easily found using 7×50 binoculars or a 6×30 finder, where it appears as a tiny patch with a rather bright central glow (the cluster's core).

## Observation

The observer is advised to use high magnifications in order to appreciate the globular nature of this concentration of stars, which is located on the extremely rich stellar background of Sagittarius. At low magnifications, M28 retains this diffuse

and bright aspect, and in consequence is of little interest: less than ten or so stars are resolved with a 115/900 reflector and magnification of 150. The brightness of the core sharply contrasts with the milky aspect of the surrounding regions. The great declination of this cluster – negative – is not conducive to optimal observations, whatever the instrument and the seeing

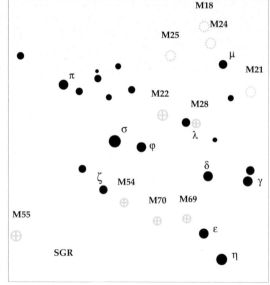

*Lost in the myriad stars in the Milky Way, M28 will be found near λ Sgr together with a multitude of all sorts of objects (open or globular clusters, dark or diffuse nebulae).*

M28. Type IV globular cluster
NGC 6626
Epoch 2000.0 coord: α: 18 h 09 m 3 s
δ: −25° 54′
Apparent ∅: 4.7′ − absolute ∅: 65 l.y.

Mv: 7.3 − Mp: 7.9 − Distance: 15 000 l.y.
Const.: Sagittarius
Discovery: Messier in July 1764
Favorable period: July in the first half of the night

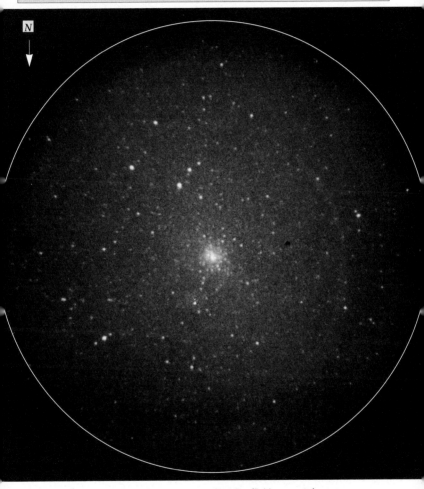

*M28 Sgr (field: see p. 82)*
(© J.-C. Merlin/Surelées Observatory).

conditions may be. The cluster is still not totally resolved with a 210-mm ×140 reflector.

# M53 and NGC 5053

## Search

The amateur can find M53 using 7×50 binoculars, 1° northeast of the star α Comae Berenices, mv 4.4). It appears as a feeble luminous patch in a region of the sky poor in bright stars.

## Observation

Out in the countryside, the cluster can be seen using a 115-mm reflector equipped with a 7-mm Or/HD eyepiece. It looks circular, the very compact core making a sharp contrast with the outer regions. With a magnification of 100, these peripheral regions appear granular, and a few stars can be isolated. On the other hand, a 210-mm reflector, ×170 magnification, will show a wonderful image: M53 reaches a diameter of almost 14′ (half that of the full Moon) with 40 or so resolved stars. The core is then reduced to a tiny and dense spot of light surrounded by a vast milky halo. M53 is one of the oldest known globular clusters, as well as one of the most remote. Almost 50 **variable stars** have been detected,

mostly of **RR Lyrae** and **Cepheid** type. M53 is 65 000 light-years away, and shines like 200 000 suns!

NGC 5053 is a pale concentration located only 1° south of M53. It appears in a 115/900 reflector as a small and diffuse star of tenth magnitude, but this cluster is a rather strange object. Its peculiarity is that it does not contain enough stars to be really classified as a globular cluster (3400 stars as against a more usual 100 000 to 500 000), nor is it an open cluster, which, on the contrary, would have stars numbered in a few hundreds.

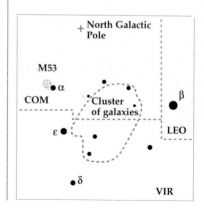

M53. Type V globular cluster
NGC 5024
Epoch 2000.0 coord: α: 13 h 12 m 9 s
δ: +18° 10′
Apparent ⌀: 10′ – absolute ⌀: 60 l.y.

Mv: 8 – Mp: 8.7
Distance: 65 000 l.y.
Const.: Coma Berenices (Berenice's Hair)
Discovery: Bode in February 1774
Favorable period: spring

N

*M53 Sgr (field: see p. 82)*
(© J.-C. Merlin/Surelées Observatory).

# M56

## History

M56 is at the limit of the Lyra and Cygnus, between the stars β Cygni and γ Lyrae. Messier discovered this formation on the 19th of January 1779, on the same night that he also discovered a comet ...

## Search and observation

Under average seeing conditions M56 is difficult to locate with a 6×30 finder, but it can easily be distinguished with 7×50 binoculars. With a 115/900 reflector and a magnification of ×70, M56 is so concentrated into a small area (2′) that the core hardly contrasts with the peripheral regions. The observer

will, however, distinguish a few stars with a 200-mm reflector, and the cluster will show an ovoidal form surrounded by a few ninth magnitude stars. The brightness of the cluster is always very uniform. High magnifications (from 1.5× the instrument's diameter) are perfectly adapted to the observation of this object. Although slight differences will be distinguished in its structure, the overall resolution will not really be improved. M56 is a rather difficult object for a novice unused to observing these faint objects.

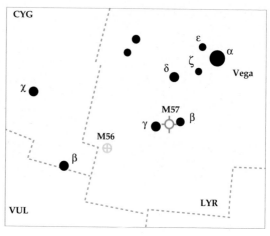

M56. Type X globular cluster
NGC 6779
Epoch 2000.0 coord: α: 19 h 16 m 5 s
δ: +30° 10'
Apparent ∅: 5.2' – absolute ∅: 60 l.y.
Mv: 8.2 – Mp: 8.8

Distance: 46 000 l.y.
Const.: Lyra (The Lyre)
Discovery: Messier the 19th of January 1779
Favorable period: July in the middle of the night

*M56 Sgr (field: see p. 82)*
(© J.-C. Merlin/Surelées Observatory).

# M92

M92 is sometimes forgotten by the astronomer, whose attention is focused on its famous and spectacular neighbor M13. Nevertheless, this globular cluster contains so many bright stars that it is almost as splendid.

## Search

At the extremity of naked-eye visibility, M92 is at the top of an equilateral triangle made together with the stars π and η Herculis.

## Observation

Using 11×80 or 12×80 binoculars, its circular form, made up of a very tight and brilliant core surrounded by a vast and diffuse halo, immediately reminds one of M13's image. A 115/900 reflector

with a magnification of 100 reveals that this group has a much denser structure than its neighbor, and is therefore less easily resolved, especially towards the center. A few scattered stars can, however, be isolated, and 30 stars are clearly visible using a 200 reflector with a magnification of 140. The disposition of the resolved stars on a milky, very granular veil gives the cluster a slightly elliptic form. M92, which has an extreme richness, has been the subject of intensive study, and

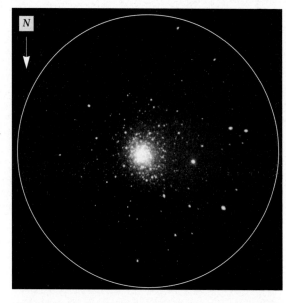

*M92 Her (field: see p. 82) (© J.-C. Merlin/ Surelées Observatory).*

M92. Type IV globular cluster
NGC 6341
Epoch 2000.0 coord: α: 17 h 17 m 1 s
            δ: +43° 09'
Apparent ⌀: 8.3' – absolute ⌀: 120 l.y.
Mv: 6.5 – Mp: 7.1

Distance: 35 000 l.y.
Const.: Hercules
Discovery: Bode in December 1777
Favorable period: July early in the night,
towards the zenith

several **RR Lyrae** type **variable stars** have been detected. As bright as 250 000 Suns and of a mass equivalent to 140 000 Suns, M92 is 15 or so billion years old. Its chemical composition is similar to that of the primordial universe.

*M92 is so concentrated that the observer needs to use high magnifications to isolate its peripheral stars. Using small diameter instruments, he or she should be resigned to seeing the stars at the center merging into a vast, milky and bright halo* (© NOAO/*Ciel et Espace*).

# PLANETARY NEBULAE

# M27

Other name **DUMBBELL**

## History

The Dumbbell Planetary Nebula, discovered by Messier on the 12th of July 1764, is one of the favorite objects for astronomers. This beautiful formation is the most extended and luminous of this type of stellar population and, with M57, is probably the most spectacular to observe.

## Search

The observer easily finds the Dumbbell using the finder, from the stars δ and γ Sagittarii, with which it forms an isosceles triangle. The novice amateur should center the star γ Sagittarii in the reticule of the finder, and should then increase the declination. γ Sagittarii and M27 are separated only by a few minutes on the same hour circle.

## Observation

M27 looks rather like an egg timer or an apple core, a shape noticeable using 11×80 binoculars. Using a 115/900 reflector equipped with a 12.5-mm focal length eyepiece, the observer will clearly distinguish the two triangular formations composing the group M27. Nuances in the tones of the nebulosity are clearly visible using average instruments. Very faint in the center, grayish tints fade towards the outer areas, mixing with the darkness of the sky until they disappear.

With apertures of at least 200 mm and under good seeing conditions, the observer can try to locate the central star, which exploded some 48 000 years ago. This **white dwarf**, of visual magnitude 13.4, is one of the brightest known stars, with a surface temperature of 85 000 °C. It possesses a companion of visual magnitude 17, at a distance of 1800 a.u. Although it appears unchanging to any observer, M27 is in fact still expanding, ejecting gas at a speed of about 30 km/s. Its apparent diameter thus increases at a rate of 1″ per century.

M27. NGC 6853
Epoch 2000.0 coord: α: 19 h 57 m 4 s
                    δ: +22° 35'
Apparent size: 8'×5'
Absolute ⌀: 2.5 l.y.

Mv: 7.6 – Mp: 7.6
Distance: 900 l.y. Mv*: 13.4
Const.: Vulpecula
Discovery: Messier in 1764
Favorable period: summer

*The Dumbbell* (© N. Sollee/ *Ciel et Espace*).

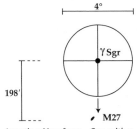

*Locating M27 from γ Sgr, with a 6×30 finder.*

*This photograph of M27 was taken using a 40-cm reflector with an image intensifier. The exposure was of only 5 seconds and the field was 30'.*
(© J.-C. Merlin/Surelées Observatory.)

# M57

Other names **THE RING NEBULA**

PLANETARY NEBULAE

The small constellation Lyra has many curiosities for amateur astronomers, the most remarkable, and thus very much observed, being The Ring Nebula.

## Search

Located between β and γ Lyra, at two fifths of the distance from β, M57's form is very distinctive. M57 is easy to locate using 7×50 or 10×50 binoculars or a 6×30 finder. It looks like a grayish round patch, in spite of its small apparent diameter (1/30 that of the Moon).

## Observation

Using a 115/900 reflector, the observer begins to distinguish the annulus, which looks like a smoke ring floating in the sky. With a magnification of 100, the annulus appears bright, with a light gray interior. The whole is sharply contrasted against the dark background sky. A 210 reflector, ×100 magnification, will show a few darker zones on the eastern and western edges of the annulus, which itself reveals its oval form, thickened at the northern and southern edges.

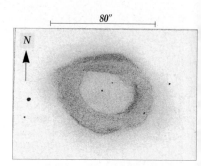

*M57 using a 355-mm reflector and a magnification of 400. The central star is a white dwarf of magnitude 14.7.* (Drawing: H. Burillier.)

M57 has in its center a **dwarf star**, which exploded some 20 000 years ago, ejecting its external layers into the surrounding space. Of magnitude 14.7, this dwarf star is 10 000 times fainter than the faintest stars visible to the naked eye. Although the dwarf star is clearly visible using a 355-mm (C14) reflector under excellent seeing conditions, the observer still needs to use an aperture of at least 300 mm. The bubble of gas which constitutes M57 is still expanding, at a speed of 30 km/s or so. This corresponds to an increase in its apparent diameter of 1″ per century.

M57. NGC 6720
Epoch 2000.0 coord: α: 18 h 53 m 6 s
                      δ: +33° 02′
Apparent size: 83″×59″
Absolute ⌀: 0.5 l.y.
Mv: 9.3 – Mp: 9.3 – Distance: 2000 l.y.

Mv*: 14.7
Const.: Lyra (The Lyre)
Discovery: Darquier in 1779
Favorable period: summer in the middle of
the night

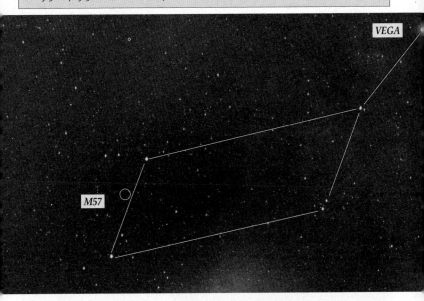

*The constellation Lyra and the
position of M57. Photograph taken
using a 135-mm lens, and an exposure time of 7 min on Tmax 400.*
(© H. Burillier/J. Motret.)

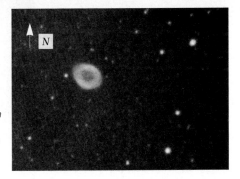

*What does M57's image look like
in a reflector? Place this photograph
3 meters from your eye ... this is
what you would see with a 200-mm
reflector.* (© J.-C. Merlin/Surelées
Observatory.)

# M76

Other name **LITTLE DUMBBELL**

## History

Méchain discovered this planetary nebula in 1780. As with most objects in the Messier Catalogue, M76 is easy to locate.

## Search

Located 50' north–northwest of the star φ Persei, at the edge of the constellation Andromeda, M76 looks like a blurred star in a 60-mm refractor with an average magnification.

## Observation

Because of its small apparent size and its irregular form, M76 requires rather high magnifications to be properly observed. With a 115/900 reflector, ×72 magnification (12.5-mm focal length Or/HD eyepiece), the observer sees its rectangular form, which is similar to that of M27. Although M76 is fainter than M27, the observation is still of interest as numerous details are visible. Two concentrations are seen using a 210-mm reflector and a magnification of 140, and a central

constriction gives M76 the appearance of a very irregular figure "8".

M76 has in its center a star of magnitude 16.6, which is unfortunately invisible to most amateur astronomers' instruments, at least under direct observation; it is obviously another matter with long-exposure photography or using CCD cameras.

*Telescopic field of M76*
(© J.-C. Merlin/Surelées Observatory).

*M76* (© CFHT/*Ciel et Espace*).

M76. NGC 650–51
Epoch 2000.0 coord: α: 01 h 41 m 9 s
                    δ: +51° 34′
Apparent size: 87″×42″
Absolute ∅: 1 l.y. – Mv: 10.1 – Mp: 12.2

Distance: 1750 to 8200 l.y. – Mv*: 16
Const.: Perseus
Discovery: Méchain in 1780
Favorable period: October and November
around midnight

# M97

Other name **THE OWL NEBULA**

## History

Discovered by Méchain in 1781, M97 is one of the most extended of planetary nebulae, as well as one of the nearest, although its actual distance is still controversial: in 1961, it was thought to be 1630 light-years away, 2380 l.y. in 1974, rising to almost 8000 l.y. according to some astronomers. An average value of 3000 l.y. has been generally adopted, corresponding to a real diameter of 3 l.y. or so.

## Search and observation

M97 is 2.5° southeast of the white star Merak (or β Ursae Majoris; see chart below), and is quite difficult to observe. The small light disk of 3′ 20″ that it makes will only be perceived under particularly good seeing conditions. It would be hopeless

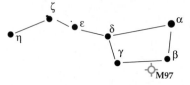

*M97 in the constellation Ursa Major (The Great Bear).*

*Observation of The Owl Nebula, a faint low-contrast object, requires perfect seeing conditions with a very clear sky. This photograph was taken 70 km from Paris, with a 106-mm fluorite refractor.*
(© C. Ichkanian/*Ciel et Espace*.)

to try to locate it when there is a Moon.

Generally speaking, with small diameter instruments (with at least 100-mm aperture), the observer will not see any detail in M97's structure,

| M97. NGC 3587 | Mv: 11 – Mp: 12 – Distance: 3000 l.y. |
|---|---|
| Epoch 2000.0 coord: α: 11 h 14 m 9 s | Mv*: 16 |
| δ: +55° 02′ | Const.: Ursa Major (The Great Bear) |
| Apparent ∅: 2.5′ – absolute ∅: 3 l.y. | Discovery: Méchain in 1781 |
| (assuming a distance of 3000 l.y.) | Favorable period: winter, spring |

*M97, or The Owl Nebula*
(© J.-M. Lecleire/F. Auchere).

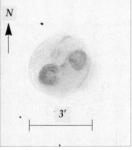

*M97 using a 200-mm reflector
(×200, OIII filter, 18 January
1995 at 00 h 40 min UT).
The central star is not visible.*
(Drawing: H. Burillier.)

apart from its apparent disk. The magnification must be sufficiently great – at least 72 times (12.5-mm focal length Or/HD eyepiece) using a 115/900 reflector – or the novice astronomer might easily overshoot the nebula while searching the region. Owners of instruments equipped with 31.75-mm diameter eyepieces (American standard) will benefit from the use of an oxygen III filter (OIII). The object will then appear sharply contrasted, standing out noticeably against the background sky. The "eyes of the owl", two dark concentrations in the middle of M97, will be visible using an aperture of at least 200 mm.

# NGC 2392

Other name **ESKIMO NEBULA**

PLANETARY NEBULAE

## History

Discovered by William Herschel in 1787, NGC 2392 is a beautiful planetary nebula, whose large apparent diameter (close to that of Jupiter) allows observations with the smallest instruments.

## Search

Under a perfectly dark sky, using 7×50 binoculars or a 6×30 finder, the observer finds the nebula half way between κ and λ Geminorum.

## Observation

With 11×80 binoculars, the nebula looks like a blurred star, while a 60-mm refractor with a magnification of 80 reveals its circular form. However, the observer needs an aperture of at least 100 mm to identify the "planetary" aspect of the nebula. Using a 115/900 reflector, the astronomer sees the central star (mv 10.5), and using a 210-mm reflector, ×200 magnification, NGC 2392 is a splendor with clearly defined edges.

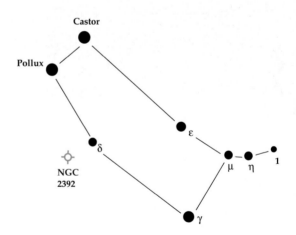

*Location of NGC 2392 in the constellation Gemini.*

NGC 2392
Epoch 2000.0 coord: α: 07 h 29 m 2 s
δ: +20° 55′
Apparent size: 47″×43″
Absolute ∅: 0.6 l.y. – Mv: 9.2 – Mp: 9.9

Distance: between 1370 and 3600 l.y.
Mv*: 10.5
Const.: Gemini (The Twins)
Discovery: Herschel in 1787
Favorable period: winter

*NGC 2392* (© S. Deconihout/*Ciel et Espace*).

NGC 2392 owes its popular name of the Eskimo Nebula to its photographic appearance: long exposure times give the illusion of a face surrounded by a furry hood.

# NGC 6826

Other name **BLINKING NEBULA**

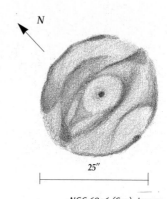

## Search and observation

Located 30′ east of the double star 16 Cygni, NGC 6826 is an object rather difficult to observe. It is in fact luminous enough to be made out with binoculars or a finder, but at least a 100-mm aperture is needed to see its form unambiguously.

With a 60-mm refractor, NGC 6826 looks like a diffuse nebular patch and is relatively difficult to see. While the nebula is luminous and compact

N

25″

*NGC 6826 (Cyg). Image of the planetary nebula, 22 June 1994 at 22 h 45 min (UT) at Puimichel, using the 1-m, f/d 3.3 reflector, ×1000 magnification and an OIII filter.*
(Drawing: J.-M. Lecleire.)

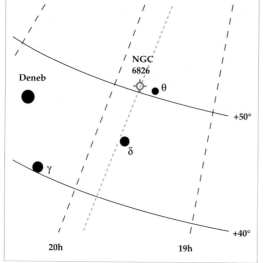

δ Cyg and NGC 6826 belong to the same hour circle (red dotted lines), which simplifies the search for the nebula: point the instrument to δ Cyg and increase the declination.

NGC 6826
Epoch 2000.0 coord: α: 19 h 44 m 8 s
                    δ: +50° 31′
Apparent ∅: 36″ – Absolute ∅: 0.6 l.y.
Mv: 8.8 – Mp: 9.8

Distance: 3500 l.y
Mv*: 11
Const.: Cygnus (The Swan)
Favorable period: late July around 00 h
towards the zenith

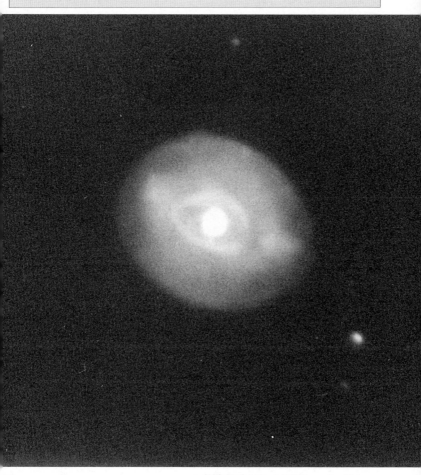

NGC 6826. (© Pic du Midi Observatory/*Ciel et Espace*.)

using a 115/900 reflector and a magnification of 100, with a magnification of 150 the central star (mv 11) "periodically" appears, as a "blinking" star. The complete formation is clearly visible using a 210-mm reflector and a magnification of 200.

# NGC 7009

Other name **THE SATURN NEBULA**

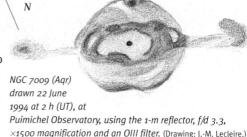

*N*

## History

William Herschel discovered this beautiful easy to observe planetary nebula in 1782.

## Search

Located 1° west of the star *v* Aquarii, at the edge of the constellation Capricornius, NGC 7009 may be found using a 6×30 finder.

*NGC 7009 (Aqr)
drawn 22 June
1994 at 2 h (UT), at
Puimichel Observatory, using the 1-m reflector, f/d 3.3,
×1500 magnification and an OIII filter.* (Drawing: J.-M. Lecleire.)

## Observation

The observer will recognize the "planetary" structure of this object

The main difficulty for the beginner is to find this planetary nebula. The ideal, for these small, diffuse objects of little contrast, is to conduct the search from the graduated circles of equatorial mounts. While this is fairly easy with large-diameter instruments, for which the graduated circles in right ascension and in declination are precise enough, it is much less so for small-diameter instruments.

Unfortunately, 115/900-type reflectors, although excellent as an introduction to observation, do not have circles precise enough. Moreover, the instability of their mounts does not make the astronomer's job easier. A simple method is to use the equatorial mount, without using the graduated circles. With the instrument in position,* the observer then points to a star located either on the same hour circle or on the same declination circle (whichever may be the case) as the object. For example, for NGC 6826 (see p. 122), the nebula is located on the same hour circle as the star δ Cygni (19 h 45 m). Center this star in the finder, then release the axis and rotate the mount towards increasing declination, until the object appears in the eyepiece's field.

* For more precise information, read the chapter devoted to this subject in *Astronomical equipment for amateurs* by M. Mobberley, Springer-Verlag.

| NGC 7009 | Distance: 3900 l.y. – Mv*: 12 |
| Epoch 2000.0 coord: α: 21 h 04 m 2s | Const.: Aquarius (The Water Carrier) |
| δ: −11° 22′ | Discovery: Herschel in 1782 |
| Apparent size: 44″×26″ | Favorable period: July and August in the |
| Absolute ∅: 0.5 l.y. – Mv: 8 – Mp: 8.3 | second half of the night |

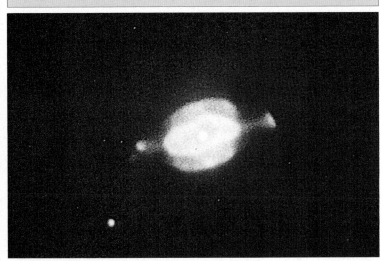

*The Saturn Nebula* (© NOAO/*Ciel et Espace*).

using a 60-mm refractor, when NGC 7009 is seen as a small round grayish patch. Using a 115/900 reflector and a magnification of 100, the nebulosity becomes more stretched out. The central zone is noticeably luminous, while the peripheral regions appear pale and attenuated. A magnification of 150 does not resolve the central star. A 200-mm aperture reveals some greenish tints, and two rings appear, from which the popular name of Saturn Nebula derives. NGC 7009 then resembles a planetary disk with two half rings.

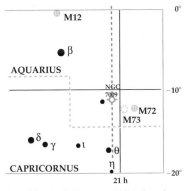

*Searching for NGC 7009, which is found within a few minutes from the star ν Cap on the same hour circle. Center this star, then slide the mount along the declination axis in the direction of the celestial equator (0°).*

# NGC 7293

Other name **THE HELIX NEBULA**

NGC 7293 is the closest of the five hundred known planetary nebulae, and has in consequence the largest apparent diameter: half that of the full Moon.

using 7×50 binoculars, and using 11×80 binoculars sees it as a small, very-low-contrast, circular patch. The circular formation is clearly

## Search

The Helix Nebula is located 8′ south of the star ν Aquarii, but because of its high negative declination (−20°), the amateur astronomer has much difficulty discovering it at latitudes half way up the northern hemisphere.

## Observation

In spite of its faintness, the observer can, however, locate it

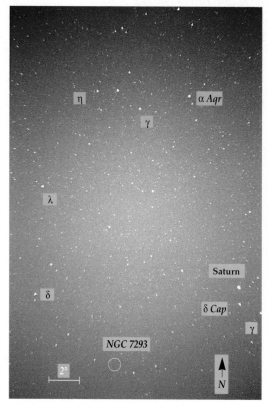

*Photograph showing the location of NGC 7293 while Saturn was near δ Cap (17 November 1993 using a 58-mm lens, with an exposure of 5 min). (© A. Merlin.)*

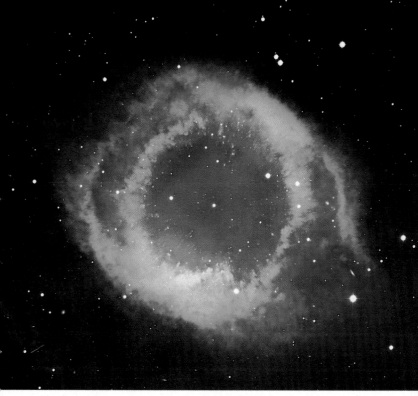

| NGC 7293 | Mv: 7.3 – Mp: 6.5 |
| Epoch 2000.0 coord: α: 22 h 29 m 6 s | Distance: 450 l.y. |
| δ: −20° 48′ | Mv*: 13 |
| Apparent size: 12′×16″ | Const.: Aquarius (The Water Carrier) |
| Absolute ⌀: 1.7 l.y. | Favorable period: autumn |

*The Helix Nebula NGC 7293* (© AAO/D. Malin/ *Ciel et Espace*).

visible using a 210-mm reflector, although it is still faint and with little contrast. The beauty of this large planetary nebula is best appreciated with photographs, where long exposure times reveal the meanders of the helix, interlaced in a superb double-ring structure. By comparison, direct observation with any instrument may disappoint the amateur, but at this point imagination can take over …

# NGC 7662

## Search

NGC 7662 is located 25′ southwest of the star 13 Andromedae and looks like a disk of almost ninth magnitude whose apparent diameter is comparable to that of the planet Saturn without its rings. Although this planetary nebula is visible in any amateur instrument, there are few who will have observed it. The relative proximity of M31, the chief object in this region of the sky, is probably the cause.

## Observation

With a 6×30 finder, NGC 7662 looks like a star, but it already appears as a small blurred patch in a 60-mm refractor with a magnification of at most 60. A 115/900 reflector with a 7-mm focal length Or/HD eyepiece reveals the oval shape of the

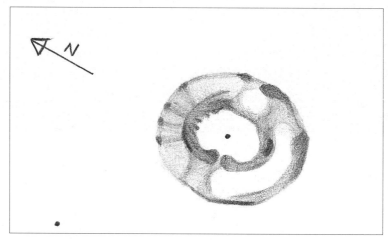

*NGC 7662 (Andromeda). Image of the planetary nebula on 24 June 1994 at 1 h 20 m (UT), using the 1-m reflector, f/d 3.3 of Pimichel, magnification of 1500, and an O3 filter.*
(Drawing: J.-M. Lecleire.)

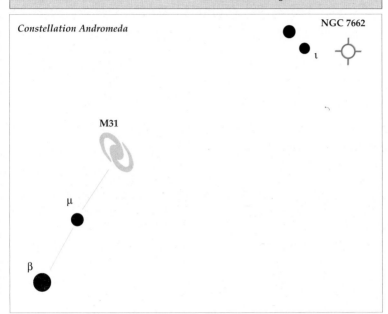

| | |
|---|---|
| NGC 7662 | Mv: 8.5 – Mp: 8.9 |
| Epoch 2000.0 coord: α: 23 h 25 m 9 s | Distance: 1800 l.y. – Mv*: 4 |
| δ: +42° 33' | Const.: Andromeda |
| Apparent size: 32"×28" | Favorable period: September and October in |
| Absolute ⌀: 20 000 a.u. | the middle of the night |

*Constellation Andromeda*

NGC 7662

M31

μ

β

*Location of NGC 7662 from M31.*

nebula, whose ring is very bright and makes a sharp contrast with the background sky. Some irregularities are visible using a 210-mm reflector, ×150 magnification. Under very good seeing conditions, NGC 7662 shows glints, bluish for some astronomers, greenish for others ...

Assiduous observers, using at least 200-mm instruments, will attempt to see the central star of the nebula. This bluish **dwarf** is a partic-

ularly hot star, of some 70 000 °C. The amateur astronomer may not find it at first, but observing as often as possible will reveal the irregular luminosity variation of this star, which is of between magnitude 12 and 16.

# DIFFUSE NEBULAE

# M8

Other name **THE LAGOON NEBULA**

M8 is one of the most beautiful diffuse nebulae we can observe. The middle of this fine gaseous structure, mixed with complex filaments, is the birthplace of new stars. Here, all the conditions for the birth of stars are met: dark clouds, **Bok globules** in the phase of gravitational collapse, **protostars**, newly born stars with a chaotic, not yet stabilized, behaviour.

## Search

In the countryside, the observer can find M8 with the naked eye despite its low position at latitudes half way up the northern hemisphere, 4° 7′ north of γ Sagittarii. A beautiful stellar cluster (NGC 6530) enveloped by a vast nebulosity, larger than the full Moon, is revealed by 11×80 binoculars.

*M8* (© NOAO/*Ciel et Espace*).

| | |
|---|---|
| M8 – NGC 6523 – Diffuse nebula | Absolute ⌀: about 50 l.y. – Mv: 6.7 |
| Associated with the open cluster NGC 6530 | Distance: 3000 to 5000 l.y. |
| Epoch 2000.0 coord: α: 18 h 03 m 8 s | Const.: Sagittarius (The Archer) |
| δ: −24° 23′ | Discovery: Le Gentil in 1747 |
| Apparent size: 45′×30′ | Favorable period: summer |

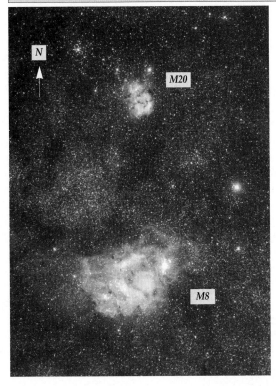

*M8 and M20 using a 230/310/707 Schmidt telescope* (© J.-M. Lopez and C. Cavadore/Pises Observatory).

## Observation

A 115/900 reflector with a 12.5-mm Or/HD eyepiece reveals the circular structure of the nebulosity, whose very bright core is separated into two distinct areas by a river of dark dust. NGC 6530, an open cluster made up of 30 or so stars, shines in the halo of glowing filaments.

Using a 210-mm reflector and a low magnification, M8 forms a compact mass in its center which rapidly vanishes into its surroundings. The dark lane, seen using the 115/900 reflector, clearly cuts through a group of luminous stars. The observer may see a few dark and misshapen structures near the formation under good seeing conditions.

**133**

# M16

## Search

Located at the edges of Sagittarius and Scutum, this nebula resembles M8 with its rounded shape, centered on a group of stars. On the other hand, M16 is not as easy to observe, and the novice may, with binoculars, scan the nebula's field without seeing it ... Whatever instrument is used, the astronomer is strongly advised to adopt a low magnification.

## Observation

Using a 115/900 reflector, M16 looks like a group of 15 or 20 stars in the middle of a faint diffuse halo, denser in its northern part. The observer needs at least a 200-mm aperture to identify the nebulous nature of this formation, and can then count over 20 stars in the middle of an extremely diaphanous veil. A suggestion of dark nebulosity can also be seen in the northern part of the nebula. The

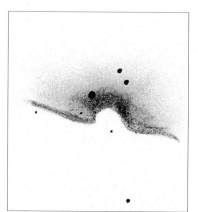 

*Left: a view of M16 using a 210-mm Newtonian telescope (×50 mag, UHC filter).*
*Right: an observation the same night using the same telescope and the same magnification, but without the UHC filter. Faint stars appear, while the nebulosity is only slightly perceptible.*
(Both drawings: H. Burillier.)

M16 – NGC 6611
Diffuse nebula associated with the open
cluster NGC 6611
Epoch 2000.0 coord: α: 18 h 18 m 8 s
                          δ: −13° 47′
Apparent size: 35′×25′

Absolute ∅: about 70 l.y. – Mv: 6.5
Distance: about 8000 l.y.
Const.: Serpens (The Serpent)
Discovery: Chéseaux in 1746
Favorable period: summer

*M16* (© ROE/AAO/D. Malin/*Ciel et Espace*).

UHC–Lumicon filter will be advantageous for those astronomers who can attach it to the eyepiece or directly to the instrument. A 200-mm reflector only shows stars up to eleventh magnitude, while on the other hand numerous details begin to appear in the filaments and other structures when the contrast is increased.

Like M8, M16, M17 and M20, M16 is the birthplace of new stars in the middle of a radiant gaseous formation that is excited by the young cluster associated with it: NGC 6611.

# M17

Other names  **THE OMEGA NEBULA**
**THE HORSESHOE NEBULA**
**THE SWAN NEBULA**

This splendid nebula is located near the intersection of the constellations Sagittarius, Serpens and Scutum.

## Search

Located 3° south of M16, M17 is a remarkable object, whatever the power of the instrument used to observe it. Found without difficulty using binoculars, this vast gaseous area looks like a stretched-out bright patch, striated with innumerable filaments interspersed with zones darkened by stellar dust.

magnification of 100 reveals an extended "2" shaped form, which looks remarkably like a swan. Its eye is a luminous star, while another one shines at the base of its neck, and with a 200×100 telescope its body is seen as a glowing veil. This superb stellar formation is enveloped by innumerable luminous filaments, and a number of nodules underline its torn and complex aspect. M17 is probably the most impressive of the diffuse nebulae that we can observe directly.

## Observation

M17 looks rather like a fan-shaped comet. Lord Rosse said it looked like the Greek letter ω and thus called it Omega, while other observers have given it the name Horseshoe or Swan. A 115/900 reflector with a

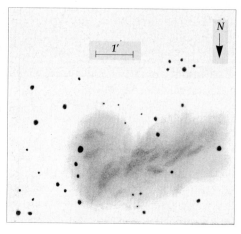

*The Omega Nebula seen with a 355-mm diameter Schmidt–Cassegrain telescope, f: 3910 mm, ×400 magnification.*
(Drawing: H. Burillier.)

M17 – NGC 6618
Epoch 2000.0 coord: α: 18 h 20 m 9 s
  δ: −16° 10′
Apparent size: 45′×35′
Absolute ∅: about 30 l.y.

Mv: 7
Distance: about 7000 l.y.
Const.: Sagittarius (The Archer)
Discovery: Chéseaux in 1746
Favorable period: summer

The Omega Nebula photographed from an observatory. When observed directly, the colors are not seen. (© ESO/*Ciel et Espace*.)

This photograph of M17, placed 4 meters from the eye, is what an observer might see using a 20-cm reflector, under good seeing conditions. The field is 30′. (© J.-C. Merlin/ Surelées Observatory.)

**137**

# M20

Other name **THE TRIFID NEBULA**

## Search

Located 1.5° north–northwest of The
Lagoon Nebula, M20 is not an easy
object to find, because it is always
very low on the southern horizon.
The instrument's power is not the
first priority, and observation of
M20 requires the clearest possible
atmosphere, an exceptional condition for objects with a declination
of –30°.

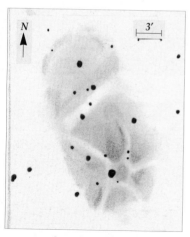

*The Trifid Nebula using a 203-mm diameter
Schmidt–Cassegrain telescope, f: 2000 mm,
×80 magnification, and a UHC filter.*
(Drawing: H. Burillier.)

Using 11×80 binoculars, the
object appears as an oval ashen
glow, with little contrast. An oblong
dark lane cuts through the nebulosity in its northeastern and north-
western regions.

## Observation

A 115/900 reflector, ×72 magnification, reveals a double star: HN 40 or
GC 24537, which is easily located at
the center of the gaseous formation.
The system actually has three
components.

| HN 40 OR GC 24537: TRIPLE SYSTEM IN THE TRIFID NEBULA | | |
|---|---|---|
| | Magnitude | Separation |
| A | 6.9 | 5' 4" |
| B | 10.6 | 10' 6" |
| C | 8.8 | |

Under good seeing conditions, the
attentive observer notices a bright
star in a bluish wispy veil north of
the nebula.

A UHC filter is perfectly adapted
to an object such as M20: the Trifid
Nebula has a pronouncedly irregular
form due to the dark lanes dividing

| M20 – NGC 6514 | Apparent size: 29′×27′ |
|---|---|
| Diffuse nebula associated with the open cluster NGC 6514 | Mv: 8.7 – Distance: 4800 l.y. |
| Epoch 2000.0 coord: α: 18 h 01 m 9 s | Const.: Sagittarius (The Archer) |
| δ: −23° 02′ | Discovery: Le Gentil in 1747 |
| | Favorable period: summer |

*M8 and M20.* (© J. Riffle/*Ciel et Espace*.)

the main lobe into three distinct branches. John Herschel first noted this peculiarity, and called it Trifid because of its clover-like shape.

# M42 AND M43

In the region located below Orion's Belt, the astronomer can easily see with the naked eye a small milky diffuse patch: this is the celebrated

M43
NGC 1982
Epoch 2000.0 coord: α: 05 h 35 m 6 s
                            δ: −05° 16′
Age: a few million years
Distance: 1500 l.y.
Absolute ∅: 30 l.y.
Apparent ∅: 20′×15′
Mv: 7
Const.: Orion (The Hunter)
Discovery: Mairan in 1731
Favorable period: winter

Orion Nebula. Binoculars will provide an exceptional view of this vast gaseous area.

## History

We owe the oldest known description to N.C. Fabri de Peiresc, in 1611. In the eighteenth century, Messier observed it as the 42nd diffuse object, and in 1880 H. Drapper obtained the first photograph. Since then, M42 has become the favorite target of amateur astrophotographers.

## Observation

M42 is a particularly dense association of dark and diffuse nebulae. The complexity of these interspersed structures, if studied with an instrument of average power (115-mm to 200-mm reflectors), will offer an unforgettable variety of color and detail. In a reasonably unpolluted sky, a 115-mm aperture instrument reveals the greenish aspect of this

*M42, a small nebulosity visible below Orion's Belt, can easily be located with the naked eye* (© H. Burillier.)

| M42 – NGC 1976 | Absolute ⌀: 30 l.y. – Max. apparent ⌀: 90′×60′ |
| --- | --- |
| Epoch 2000.0 coord: α: 05 h 35 m | Const.: Orion (The Hunter) |
| δ: −05° 23′ | Discovery: attributed to Fabri de Peiresc in 1611 |
| Age: a few million years | |
| Distance: 1500 l.y. | Favorable period: winter |

object. A UHC or Hα filter, centered on the excited hydrogen line (emission at 656.3 nanometers) will facilitate observation: surrounding stars will appear fainter, while the background sky, suddenly darkened, will increase the contrast of the nebula.

The observer must give priority to field of view, which should be as wide as possible, as M42 is a particularly spread out object: at 60′×66′, it is twice the diameter of the full Moon!

Like all diffuse nebulae, M42 is composed of gas, mostly hydrogen (90%). Observations made by the astronomical satellite IRAS in 1983 revealed the presence of cold gas, with in some regions gas in the phase of contraction and heating, a necessary condition for the birth of new stars. The Orion

Nebula as well as regions near the Horse Head are the birthplaces of new stars, which in their embryo state are called **protostars**.

M42 reveals in its center an opaque area with a small trapezoidal cluster. Its stars are separated from each other by 10″ to 20″ and can be observed using a 60-mm refractor. Very young (one or so million years) and very hot, they are in part responsible for the brightness of the surrounding nebula.

M43 is the northern part of the Great Orion Nebula; it reveals its greenish color in a 115-mm instrument. The main star, buried in the nebula, bathes it in a bluish luminous glow.

*M42 and M43. Photographed 26 December 1992, using the 520-mm telescope at Puimichel Observatory, with an exposure time of 10 min.*
(© J.-M. Lecleire.)

# NGC 2024 AND NGC 2023

## Search

NGC 2024 is an important diffuse nebula, whose apparent diameter is nearly that of the full Moon. This huge gaseous area is thus perceptible with average-sized instruments.

ξ Orionis, a splendid and brilliant 1.8 magnitude star, is separated from NGC 2024 by only 15′; its brightness is a considerable hindrance to the detection of the complex shapes of light and shade within the nebula.

### Darkening the background field

A UHC or Hβ filter is therefore recommended, since they noticeably reduce the brightness of the star ξ Orionis. In addition, keeping the nebula towards the edge of the field of view helps to locate it, since ξ and its bright glow are then outside the field of view.

## Observation

With a 115/900 reflector and a small magnification of 50 (Or/HD 18 mm), NGC 2024 reveals a circular structure which seems to be cut into two zones by a minute dark strip, perfectly visible with a 200-mm reflector. The nebulosity is remarkable and shows many irregularities at its edges.

*NGC 2024*
(© J.-M. Lecleire).

| NGC 2024 | Distance: about 1 300 l.y. |
|---|---|
| Epoch 2000.0 coord: α: 05 h 41 m 9 s | Const.: Orion |
| δ: −01° 50' | Discovery: Herschel in 1746 |
| Apparent ⌀: 30' | Favorable period: winter |

*NGC 2024 and the Horse Head* (© J. Riffle/*Ciel et Espace*).

Moving 20' southward, the observer sees HD 37903, an 8.5 magnitude star, and in the middle of a faint bluish halo NGC 2023 reveals itself. Its circular nature gives it the appearance of a blurred star.

# NGC 7000

Other name **The North America Nebula**

DIFFUSE NEBULAE

NGC 7000 is a chaotic complex within which dark veils of absorbing matter are mixed with vast stellar clouds excited by an extremely hot blue star: HD 199579. Because of its shape, M. Wolf named it The North America Nebula in 1890.

## Search

The vast nebulosity, which spreads over 1.5°, is located under 3° east of the star α Cygni, or Deneb.

## Observation

Using a 115/900 reflector, the observer should choose to use a 40-mm AH eyepiece, since it has the longest focal length of 24.5-mm diameter eyepieces, and therefore provides the lowest magnification. On a 115/900 reflector, the resulting magnification of 22.5, for an absolute field of 1.5°,

allows the astronomer to see the increased brightness of NGC 7000 compared with the rich stellar background of the Milky Way. The astronomer will obtain still greater contrast with a 25-mm Or/HD eyepiece and a red filter (W29), which, although more suitable for photographic use, will nicely enhance the darkness of the background sky. The most impressive image of NGC 7000 will, however, be obtained using binoculars alone. The analogy with the shape of North America is really striking using 11×80 binoculars under a perfectly pure and dark sky.

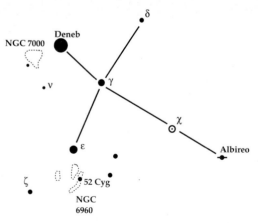

*Location of NGC 7000. This nebula can be located with binoculars near Deneb, but only in a perfect sky, without light pollution ...*

| Diffuse nebula associated with the star HD 199579 | Absolute ∅: 45 l.y. – Mv: 5 |
|---|---|
| Epoch 2000.0 coord: α: 20 h 58 m 8 s | Distance: about 3000 l.y. |
| δ: +44° 20′ | Const.: Cygnus (The Swan) |
| | Discovery: Herschel in 1786 |
| Apparent size: 120′×100′ | Favorable period: July and August |

... or this kind of intense lighting, unregulated and badly managed, which totally precludes any astronomical observation.
(© J.-M. Lecleire.)

NGC 7000 using a 250/310/707 Schmidt telescope
(© J.-M. Lopez and C. Cavadore/Pises Observatory.)

# GALAXIES

# M31, M32, AND NGC 205

## Search

The milky cloud M31 is the most remote object we can observe with the naked eye. This immense island universe covers an area the diameter of five full Moons! M31 is accompanied by two dwarf satellite galaxies, each visible using binoculars: NGC 205 and NGC 221 (M32).

## Observation

Only binoculars will give a complete panorama of M31, which, bulging at its center with a resplendent shining core surrounded by a vast halo, is an exceptional sight. Using small diameter instruments, most novice astronomers are disappointed, as they do not always appreciate what the image contains. There are no clouds of dark matter, as in photographs, and the core, just distinguishable within the gigantic halo that encloses it, shades into its surroundings. There is no star, no detail, only an elongated bulging shape with a central bulge. This is all one sees of the Great Andromeda Galaxy with a small instrument. The idea of photons travelling for two million years from beyond our galaxy into the living eye ... this is what makes M31 so extraordinary and exciting.

NGC 205, an oval with a very luminous core contrasting sharply with the halo, appears west of M31.

South of M31 and almost enclosed in its gigantic arms, the circular form of M32 (NGC 221) is easily located.

The observer needs an instrument of at least 200 mm for detailed observations of the Andromeda

---

M32
NGC 221
Type E2 galaxy
Epoch 2000.0 coord: α: 00 h 42 m 7 s
δ: +40° 52′
Apparent ∅: 14′
Distance: 2.2 million l.y.
Const.: Andromeda
Discovery: Le Gentil in 1749
Favorable period: autumn

---

NGC 205
Type E6 galaxy
Epoch 2000.0 coord: α: 00 h 40 m 3 s
δ: +41° 41′
Apparent size: 8′×3′
Mv: 4
Distance: 2.2 million l.y.
Const.: Andromeda
Discovery: Messier in 1773
Favorable period: autumn

| | |
|---|---|
| M31 – NGC 224 | Mv: 4 |
| Type Sb galaxy | Distance: 2.1 million l.y. |
| Epoch 2000.0 coord: α: 00 h 42 m 7 s | Const.: Andromeda |
| δ: +41° 16' | Discovery: Al Sûfi around 950 |
| Apparent size: 150'×50' | Favorable period: autumn |

½° Full Moon

θ And

M32    M31

NGC 205

*M31 using a 250/310/707 Schmidt telescope*
(© J.-M. Lopez and A.N. Jacquey/Pises Observatory).

Galaxy and its companions. When observed in the countryside with a dark sky, between two and four dark lanes, coiled within the spiral arms, are particularly noticeable in the western and southeastern parts of the galaxy.

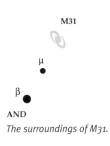

NGC 7662    ο

ι

M31

μ

β

AND

*The surroundings of M31.*

# M33

Other name **THE TRIANGULUM GALAXY**

South of Andromeda, the small Triangulum Galaxy becomes visible as soon as darkness falls in early winter. M33 is the center of interest of this region of the sky. This superb spiral galaxy, seen face on, belongs to the **Local Group**, and is our closest neighboring galaxy after **The Magellanic Clouds** and M31. Nevertheless, in spite of its large apparent size, M33, of magnitude 6.7, is not an easy object to observe as it is too spread out and requires exceptional atmospheric conditions. A dark sky with no ambient light is vital in order to see M33's pale nebular glow, which would otherwise be totally lost against its background.

## Search

Using 10×50 binoculars, the observer can locate M33 on the segment made by β Andromedae and α Tauri (Aldebaran) at the center of a triangle made by three stars of eighth magnitude. A 115/900 reflector with a low magnification reveals the galaxy, which appears as a slightly stretched shape with little contrast. A tiny point of light in its center denotes the presence of a faint core.

## Observation

Using an instrument of average diameter (150–200 mm), the image is unexceptional, while with a 200/2000 telescope, ×80 magnification, one of the arms stretches from the core towards the periphery, before terminating at a tenth

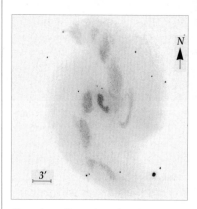

*M33 using a Newtonian telescope, 210-mm diameter, f 1260 mm and a magnification of 40 [×84]. (Drawing: H. Burillier.)*

M33 – NGC 598
Type Sc spiral galaxy
Epoch 2000.0 coord: α: 01 h 33 m 9 s
δ: +30° 39'
Apparent size: 60'×35'

Mv: 5.3
Distance: 2.5 million l.y.
Const.: Triangulum (The Triangle)
Discovery: Messier in 1764
Favorable period: autumn

*M33 using a 250/310/707
Schmidt telescope.*
(© J.-M. Lopez and C. Cavadore/
Pises Observatory.)

*The low surface bright-
ness of M33 makes it hard
to find and to observe.
This photograph shows
the low contrast between
the galaxy and the back-
ground sky. It is compar-
able to the image seen
using 7×50 binoculars
on a perfectly clear night.*
(© J.-C. Merlin/Surelées
Observatory.)

magnitude star. At this extremity,
NGC 604, a small diffuse nebula
belonging to M33, is clearly visible,
and looks like a small blurred star.
Observers can see it without difficulty using a 115-mm reflector.

# M51 AND NGC 5194–95

Other name **CANES VENATICI GALAXY**

## Search

Located 3° southwest of η Ursae
Majoris, M51 is one of the most
spectacular galaxies in the Messier
catalogue. The beginner should,
however, be aware of the perhaps
familiar photographic image, since,
as with any spiral galaxy, M51's
visual aspect is rather different.
The photograph on the left shows
M51 as it would appear using an
instrument of 150 to 200 mm, under
excellent seeing conditions.

*This photograph, when placed 2.5 m away,
shows M51 as it would appear using a
130-mm diameter instrument.* (© J.-C. Merlin/
Surelées Observatory.)

The observer can discover this
galaxy, seen face on, using binoculars. Once again, the sky must
be free from any light pollution,
including that of the Moon, since
M51 is so spread out that it easily
disappears against the background
luminosity.

## Observation

With a 115/900 reflector and a magnification of 50, the galaxy, large
and with little contrast, looks like
two blurred but relatively bright
"stars" lost in a vast mist. Under a
perfectly dark and stable sky, the
observer can see a spiral.

NGC 5195 is an irregular galaxy
physically linked to NGC 5194. It is
clearly visible with small-aperture
instruments, and its chaotic structure resembles that of M82. A spiral
arm can be seen using instruments
of at least 200-mm diameter, at low
magnification. A tenuous filament,
joining the two galaxies, can just
be seen using a 115/900 reflector.
Southwest of the formation, the
observer also sees a luminous

M51 – NGC 5194–95
Type Sc spiral galaxy
Epoch 2000.0 coord: α: 13 h 29 m 9 s
                    δ: +47° 12′
Apparent size: 11′×7.8′

Mv: 5.3 – Distance: 22 million l.y.
Const.: Canes Venatici (The Hunting Dogs)
Discovery: Messier in 1773
Favorable period: 15 April at midnight, at the zenith

*M51* (© CFHT/*Ciel et Espace*).

area, which is in fact another arm of the galaxy.

But, once again, the results obtainable are more a function of the conditions of observation (transparency of the atmosphere, extraneous light sources, turbulence, etc.) than of the power of the instrument.

# M64

Other name **THE BLACK-EYE GALAXY**

M64 is not often targeted by novice astronomers, since there are few luminous stars in this region of the sky to help in its location.

## Search

M64 is, however, an interesting object, located 55′ northeast of the star 35 Comae Berenices and 5° south of the north galactic pole. It is so bright that the observer can easily find it using binoculars.

## Observation

A 115-mm reflector with a small magnification straight away reveals M64's milky ovoid form, so perfect that it resembles an egg; the core shines brightly in its center. Using an aperture from 200 mm, the astronomer notices a dust lane to the north of the galaxy, which is probably the birthplace of new stars. A further patch completes the structure which gives this galaxy the name Black-eye. This can be very clearly seen when the eye becomes used to the darkness.

*M64* (© CFA/Schild/*Ciel et Espace*).

*M64 using a 400-mm reflector, with an image intensifier and an exposure time of 5 s, on Tmax 400 film.* (© J.-C. Merlin/Surelées Observatory.)

M64 – NGC 4826
Type Sb spiral galaxy
Epoch 2000.0 coord: α: 12 h 56 m 7 s
δ: +21° 41′
Apparent size: 9.2′×4.6′

Mv: 8.5
Distance: 25–50 million l.y.
Const.: Coma Berenices (Berenice's Hair)
Discovery: Bode in 1779
Favorable period: spring

# M65 AND M66

Because of its position near the north galactic pole, the constellation Leo can be observed without having to look through the Milky Way. Using small-diameter instruments, the astronomer can identify and observe innumerable galaxies and can locate 40 or so of them, among which are the couple M65 and M66, which can be found with a 115/900 reflector.

## History

Pierre Méchain discovered these galaxies in March 1780, without, however, paying much attention to them. Their unchanging position simply confirmed that these small diffuse patches were not the comets which were given so much attention in those days ...

## Observation

Using a wide-angle eyepiece, the observer sees both galaxies in the same field of view. M66 is almost face-on, while its companion M65 is seen more obliquely. Using a 60-mm refractor and a magnification of 60 or so, the elliptical shape of M65 is immediately revealed, while a 115/900 reflector shows the bright core of the galaxy. Using an aperture of at least 200 mm, the colors and textures of M65 remind the astronomer of those of M82.

M66, a beautiful spiral galaxy looking like a slightly oval nebula, lies 20' southeast of M65. Its

*M65* (© AAO/D. Malin/*Ciel et Espace*).

M66
Type Sb spiral galaxy
Epoch 2000.0 Coord: α: 11 h 17 m 6 s
δ: +13° 17'
Apparent size: 7.8'×3.6'
Mv: 9.7
Distance: 38 million l.y.
Const.: Leo (The Lion)
Discovery: Méchain, at the same time as M65
Favorable period: spring

M65

| | |
|---|---|
| M65 | Mv: 9.5 – Distance: 38 million l.y. |
| Type Sb spiral galaxy | Const.: Leo (The Lion) |
| Epoch 2000.0 coord: α: 11 h 16 m 3 s | Discovery: Méchain in 1780 |
| δ: +13° 23′ | Favorable period: spring |
| Apparent size: 8.1′×2.5′ | |

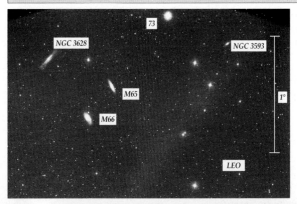

*M65 and M66, using a 250/310/707 Schmidt telescope* (© J.-M. Lopez and N. Mont/Pises Observatory).

*M66* (© AAO/D. Malin/*Ciel et Espace*).

extremely luminous core, more extensive than that of M65, is revealed with a 115/900 reflector, whereas a 200-mm reflector with an average magnification shows a dark area starting near the center and extending to the periphery of the galactic disk.

# M81

M81 is a splendid galaxy visible with any instrument.

## Search

M81 looks like a blurred star using a 6×30 finder. The observer will find it more easily with binoculars: from α Ursae Majoris and λ Draconis, extend an isosceles triangle to the west (at the passage at the superior meridian); the pair of galaxies M81 and M82 will appear. Finding M81 is easier with an equatorial mount: point towards λ Draconis (fifth magnitude), increase the right ascension, and M81 will soon appear. Using a 115/900 reflector, a 25-mm Or/HD

M81, M82 (© J.-M. Lopez and C. Cavadore/Pises Observatory).

*In the spring of 1993, a supernova appeared in the galaxy M81. Discovered by a Spanish amateur astronomer, SN 1993 J was visible for several weeks before disappearing (photograph taken on Tmax 400 film using an image intensifier and an exposure of 5 s).*
(© J.-C. Merlin.)

| M81 – NGC 3031 | Mv: 6.9 |
| --- | --- |
| Type Sb spiral galaxy | Distance: 8.5 million l.y. |
| Epoch 2000.0 coord: α: 09 h 55 m 6 s | Const.: Ursa Major (The Great Bear) |
| δ: +69° 04′ | Discovery: Bode in 1774 |
| Apparent size: 24′×13′ | Favorable period: winter and spring |

*The galaxy pair M81 and M82 shown as it would be seen using a reflector of about 200-mm diameter (© A. Fujii/Ciel et Espace).*

eyepiece is advised, because of its low magnification (×36) and its wide field of view (65′ or so).

## Observation

The observer can see a vast halo with a bright core at its center. Using a 210-mm reflector and a magnification of 140, the careful astronomer sees that the core is not a uniform mass, but is made up of three distinct grainy formations. With an instrument of at least 200 mm, the start of a spiral arm can be seen in the southern part of M81, near two stars of twelfth magnitude at the galaxy's edge.

# M82

## History

J.E. Bode probably did not realise that he had discovered one of the most fascinating objects of modern astronomy when he first observed this nebulosity in December 1774. That same year, Messier catalogued M82 with this comment: "... a feeble and elongated nebula."

## Search

Located 38′ north of M81 on the same hour circle, M82 is seen edge-on, its elongated luminous shape extending for over 10′, making it an easy binocular object.

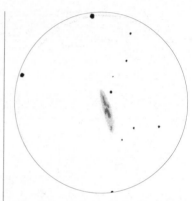

*M82 using a 200-mm reflector, f: 10, 26 and 12.5-mm eyepiece, magnification of 77 and 160.* (Drawing: H. Burillier.)

*M82 as seen using a 150- to 200-mm diameter instrument.* (© J.-C. Merlin/Surelées Observatory.)

| M82 | Mv: 8.4 |
| --- | --- |
| Type Sb spiral galaxy | Distance: 11 million l.y. |
| Epoch 2000.0 coord: α: 09 h 51 m 9 s | Const.: Ursa Major (The Great Bear) |
| δ: +69° 56′ | Discovery: Bode in 1774 |
| Apparent size: 13′×4′ | Favorable period: winter and spring |

*The galaxy pair M81 and M82 is shown here as it would be seen using a reflector of about 200-mm diameter*
(© A. Fujii/*Ciel et Espace*).

## Observation

Its irregular form, typical of such galaxies, is easily seen using instruments of at least 100-mm diameter. M82 is the prototype of chaotic galaxies whose stars are not resolved, and it presents a typically irregular aspect. The continuous activity in M82 has occupied astronomers for many years. Some theoretical models predict the presence of a very energetic and mysterious object at the center of the galaxy which could be the origin of this activity, and which might perhaps be a giant black hole …

From M81, M82 can be found simply by moving the declination axis of the equatorial mount northwards. M82 appears at the edge of the field at the same time that M81 disappears on the opposite side. Using a low magnification and a wide field eyepiece, the amateur astronomer can see both galaxies in the same field of view.

# M94

M94 is a spiral galaxy visible with any instrument. The observer can even see its nebular shape using a 60-mm refractor.

## Search

This splendid formation, which is very compact and edge-on, makes an isosceles triangle with the stars α and β Canum Venaticorum. The novice astronomer can locate it very easily using binoculars or a finder. It is, however, advisable to change the original 5×24 finder on most 60-mm refractors and 115-mm reflectors for a 6×30 finder, which will be easier to use and give greater luminosity.

## Observation

Using a 115/900 reflector, M94 shows a very bright core, surrounded

by a faint diffuse area, and resembles an unresolved globular cluster: a circular object without visible detail. However, the observer notices that M94 is sharply contrasted against the surrounding background sky, which is particularly empty and

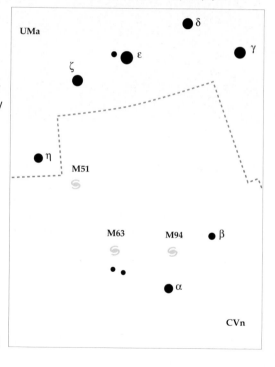

M94 – NGC 4736
Type Sb spiral galaxy
Epoch 2000.0 coord: α: 12 h 50 m 9 s
              δ: +41° 07′
Apparent size: 13′×11′

Mv: 8.2 – Distance: 20 million l.y.
Const.: Canes Venatici (The Hunting Dogs)
Discovery: Méchain in 1781
Favorable period: March and April in the mid-dle of the night

*M94* (© NOAO/*Ciel et Espace*).

dark. A 200-mm reflector will not reveal any more detail: the form of the galaxy will appear more oval and the halo more stretched than with a 115 reflector, but still no detail will be visible in the core or in the galactic halo. This is probably because of the relative brightness of M94's center, which obliterates any detail further out.

# M101

## Search

Located 5° 30′ east of Mizar, M101 is a vast face-on spiral galaxy. Although its apparent diameter is almost that of the full Moon, it is very difficult to locate. M101 is faint and very spread out, so much so that perfect seeing conditions are necessary and there is no chance of finding it in urban areas. Using a 115/900 reflector, the astronomer needs an average magnification of about 70, as too low a magnification would not provide enough contrast between the faint halo of the galaxy and the background sky. M101 could then actually be within the field of view but effectively invisible.

## Observation

In an instrument of average diameter,

M101 appears as a diffuse sphere and shows no detail. A 210-mm reflector reveals little more, apart from a tiny core only just visible. All the same, the amateur astronomer should not neglect this object, as it has been shown to be rich in appearances of **supernovae**.
A very consistent study of its field could give the careful and patient observer a surprise ...

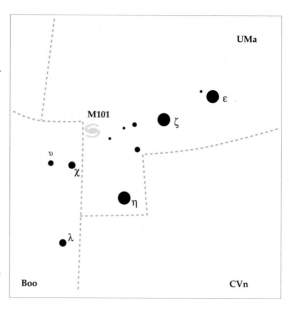

M101 – NGC 5457
Type Sc spiral galaxy
Epoch 2000.0 coord: α: 14 h 03 m 2 s
                     δ: +54° 21′
Apparent ⌀: 26′

Mv: 7.9
Distance: 17 million l.y.
Const.: Ursa Major (The Great Bear)
Discovery: Méchain in 1781
Favorable period: spring

M101 (© A. Fujii/*Ciel et Espace*).

# M104

Other names **SOMBRERO HAT GALAXY**
**SOMBRERO GALAXY**

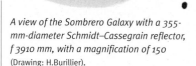

M104 is one of the greatest treasures
to contemplate in the whole of
astronomy. This splendid spiral
galaxy, which looks rather like
Saturn, fascinates most amateur
astronomers, and even more so
those who have developed a passion
for astrophotography.

*A view of the Sombrero Galaxy with a 355-
mm-diameter Schmidt–Cassegrain reflector,
f 3910 mm, with a magnification of 150*
(Drawing: H.Burillier).

## History

Although Charles Messier probably
never observed this object, which
was discovered by Pierre Méchain in
1781, Camille Flammarion proposed in
1821 that this vast galaxy should be
called M104 from the famous Messier
catalogue. The request was adopted,
and later generally recognized.

## Search

Located 45′ west of Spica (α Virgo),
M104 is easily recognizable with
7×50 binoculars. Observing the
isosceles triangle that it forms with
the stars Spica and γ Virginis helps
to locate it. A pointing can be made
quickly from γ Virginis on any equatorial mount instrument. Once the
star is spotted, simply decrease the
coordinate of the declination axis by

10° or so. M104 then appears in the
field of view.

## Observation

A 115-mm reflector with a low magnification reveals M104 as an elongated form, with a large bulge at its
center. The galactic core, which is
extremely luminous, lies in a vast,
diffuse, circular halo.

Using a 200-mm aperture instrument and in favorable conditions, a
dark lane of dust is discernible. This
strip bisects the galaxy's bulge,
whose northern part is even more
luminous. As soon as this dark strip
of matter (which in reality is an edge-
on spiral arm) can be distinguished,
the contrast with the southern part is
indeed seen to be greater.

M104 – NGC5494
Type Sb spiral galaxy
Epoch 2000.0 coord: α: 14 h 40 m 0 s
δ: −11° 37′
Apparent size: 7.1′×4.4′

Mv: 8.0 – Distance: 40 million l.y.
Const.: Virgo
Discovery: Méchain in 1781
Favorable period: March in the middle of the night

*Sombrero galaxy (M104)*
(© ESO/*Ciel et Espace*).

*M104 with a 250-mm f/4.5 reflector and an exposure time of 45 min* (© B.Gaillard/ Pises Observatory).

# OTHER OBJECTS

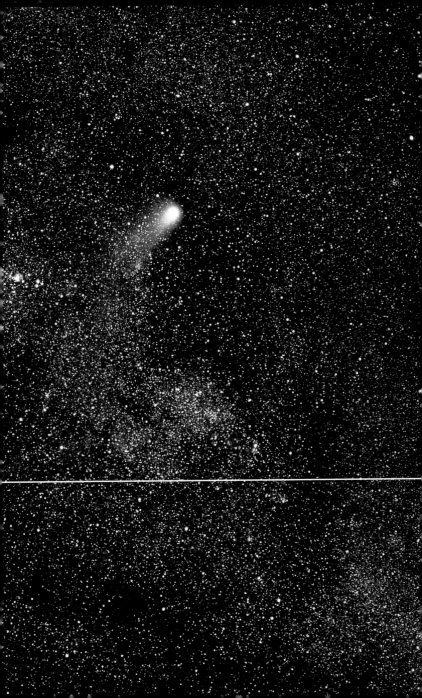

# METEORS

Because of its orbit around the Sun, the Earth is continuously bombarded with extraterrestrial dust of various sizes. The term meteor describes the luminous phenomenon we observe when a cosmic particle enters the Earth's atmosphere. The appearance of a so-called shooting star can be sporadic, or it may come from a shower whose origin (the radiant) is localized.

Russian postage stamps, showing represen-tations of bolides. (© Collection M. Verdenet.)

This meteor from the Perseids shower was photographed on the 12th of August 1992 at 22 h 05 min UT (the camera, using an f 1.8, 50-mm lens and an exposure time of 15 min, was mounted on a tripod).
(© Alpha Centauri Club, Carcassonne, G.L. Carrié.)

| Characteristics of meteors | Average temperature during ablation: |
|---|---|
| Average speed of entry into the atmosphere: 40 km/s | 3000 °C |
| | Magnitude: |
| **Ablation** altitude: begins at 140 km or so and ends at around 100 km | Mv 0 for a meteor lighter than 1 g |
| | Mv −4 for a bolide of about 50 g |

*Meteors originating from the Perseids shower. The small blurred patch at the top right is M31.*
(© Y. Yanacushi/ *Ciel et Espace.*)

Meteor showers are of cometary origin. When they approach the Sun, comets eject gas and vast clouds of matter behind them (the tail), some of which periodically cross the Earth's orbit. For example the Perseids, visible every summer, originate from the Comet Swift–Tuttle.

## Observation

Some meteors are brighter than others; those whose magnitude is estimated at under 4 (i.e. brighter than Venus) are called bolides. Their appearance is sometimes accompanied by an explosive sound or sonic boom not unlike the flapping of a bird's wing. Some bolides survive the passage through the Earth's atmosphere and fall to the ground; they are then called meteorites.

The observation of meteors is more than easy: you only need your eyes and ... maybe a deckchair: do not forget that visual acuity also depends on the comfort of the observer! The astronomer should observe between 40° and 70°, where more meteors appear than at the zenith. On the chart, the observer should carefully note the direction of the meteors in relation to the stars, the estimated magnitude, the apparent velocity, the time and the order of their appearance. The data can be used for statistical studies, and may have scientific value. Nowadays, visual observation is the basis of meteoritic astronomy ...

# COMETS

From antiquity people have thought that comets, called "dust clouds" by the Chaldeans, have prophetic powers, especially in the foretelling of various catastrophes. Chinese annals record them from 613 BC. Today, astronomers make intensive study of these bodies from the outermost bounds of the solar system in the hope of extracting their secrets, in particular clues to the age of the universe.

## Observation

With a reflector, a comet is seen to be made up of three parts. The core, composed of frozen matter, gas, rocks and metals, is the brightest part and measures a few kilometers across. The coma is the nebulous matter surrounding the core, and has a radius varying between 50 and 100 000 km, and sometimes more. The tail comes from the **sublimation** of the ice of which the core is composed. It develops as the comet approaches the Sun and can spread out over millions of kilometers.

Amateur astronomers play a particularly important part in the observation of comets: one comet in three is discovered by a non-professional!

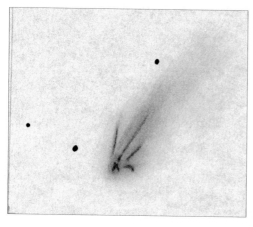

*Comet Halley, seen from France on 7 May 1986 at 22 h 10 min UT, using a 115/900 reflector (H 20-mm eyepiece, ×45 magnification).* (Drawing: H. Burillier.)

*Comet Swift–Tuttle, 24 November 1992 at 17 h 43 min UT (using a 160-mm Schmidt telescope, 20 min exposure on a TP 2415 hypered film).* (© J.M. Lopez/ J.P. Sombart/Pises Observatory.)

Direct observation using an eyepiece still has a future: drawings of their shapes, their positions relative to surrounding stars, evaluations of magnitude, trajectories, etc. are all motivations for the beginner, whatever the power of the instrument.

Why not become a comet hunter? Steady and meticulous observation of large areas of the sky, night after night, can bring luck to the patient observer …

**173**

# DOUBLE STARS

The Sun is a lone star; however, most stars are accompanied by a companion, and sometimes two or more. A double star whose components can be separated with an instrument is called a visual binary. Over time, the observer can see the movements of the bodies along their orbits, can measure their angular positions (θ), the distance between their components (ρ) and can determine their periods of revolution.

## History

In 1802, William Herschel was the first to suggest the existence of physically linked binary systems made up of two stars orbiting around a common center.

## Observation

The observation of double stars just requires little equipment and a large number can be found using a 60-mm refractor. The use of a large diameter reflector does not favor the observation of these objects: they are too sensitive to atmospheric turbulence and too bright, and do not attain their theoretical

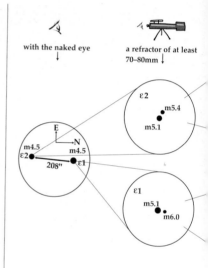

with the naked eye ↓          a refractor of at least 70–80mm ↓

resolution. However, 80–100-mm refractors are excellent, and the observer should use a high magnification eyepiece in order to separate very close components.

## Using a refractor and a reflector

Albireo, or β Cygni, one of the most colorful binaries, was identified as a

| Name | Magnitudes | | Separation, ρ | Constellation |
|------|------|------|------|------|
| | m1 | m2 | | |
| Alcor and Mizar | 2.1 | 4.0 | 708″ | Great Bear |
| α1 and α2 Cap | 3.6 | 4.2 | 376″ | Capricornus |
| θ1 and θ2 Tau | 3.5 | 4.0 | 337″ | Taurus |
| ε1 and ε2 Lyr | 4.5 | 4.5 | 208″ | Lyra |

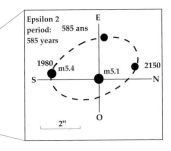

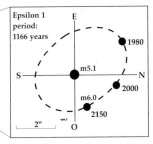

Mayer identified Gamma Andromeda as a binary system in 1777, while in 1842 Sturve found a triple component. Only the first two components are resolved with small instruments. This pair is of a rare beauty, with a yellow–orange primary star and an emerald-green secondary star.

m1: 3.0
m2: 5.1
θ: 63°
ρ: 10.01″

Epsilon Boötis, identified as a double star by Herschel in 1779, is famous for its diversely colored components. This very bright binary system is not always easy to resolve, and a 70-mm refractor with a high magnification is required. The primary star is yellow, while the secondary is bluish or greenish in color.

double star in 1755 by Bradley. The observer can resolve it using a 60-mm refractor, and can see its gold and greenish-blue tints.

m1: 3.0
m2: 5.3
θ: 54°
ρ: 34.3″

m1: 3.0
m2: 6.3
θ: 338°
ρ: 2.79″

# VARIABLE STARS: MIRA CETI

## History

In northern Germany, on the 13th of August 1596, the Lutheran pastor and amateur astronomer Fabricius observed a third magnitude star which had not been mentioned in any catalog; he called it "Marvellous star in the whale's neck" (Mira Ceti). In October the same year, the star had disappeared. However, the astronomer Bayer noted it in his atlas in 1603, and in 1638 Holwarda found it again during a lunar eclipse, losing it in the summer only to find it again on the 7th of November. He immediately concluded that the luminosity of

0214-03 Mira Cei (Omi Cet)
1900: 02h 14m 18s -03° 26.1′
1950: 02h 16m 49s -03° 12.2′
2000: 02h 19m 21s -02° 58.3′
Mira – mv 2.0 to 10.1 – per: 331.6 d – sp M5e–M9e

annual precession
+ 3.03 s + 0.278′
eq. 1900.0

*Location charts for Mira Ceti. Numbers indicate the magnitudes of comparison stars. Chart B is reversed since it represents the field of observation as seen from a reflector. (© AFOEV.)*

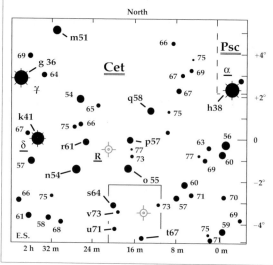

a: α Tau 1.3
b: α Gem 1.7
c: δ Gem 1.9
d: α Ari 2.2
l: μ Cet 4.6
e: α Cet 2.7
f: δ Ari 3.0

Eq. 1900

A

| o Cet | Mv: 2.0 to 10.1 |
| Mira Ceti | Period: 331.6 days |
| Epoch 1950 coord: α: 02 h 14 m 18 s | **Spectra**: M5e–M9e |
| δ: −03° 26.1′ | Const.: Cetus (The Whale) |
| Type: Mira Ceti | Favorable period: autumn |

certain stars changed over time, and the story of variable stars began ...

## Observation

Long period variable stars (of Mira type) are numerous, and the *General Catalogue of Variable Stars* lists over 6000 of them. Luminosity amplitudes vary between 2.5 and 7 magnitude, and periods between 80 and 1000 days. They are fascinating to study, and can be seen using binoculars or any small instrument.

The most remarkable variable star of this group is χ Cygni, which fluctuates between magnitude 3 and 14.5 in just over 400 days. Between these two extremes, the visual brightness of this star varies by a factor of 10 000!

Mira Ceti, the eponymous long period variable star, is close to

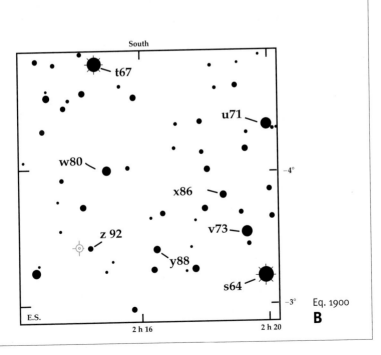

**177**

second magnitude at maximum, and tenth magnitude at minimum brightness, with a cycle of 331.6 days.

## Visual estimation of stellar magnitude

Observers of variable stars visually determine magnitude using the method defined by Argelander in 1840, which, with some experience, is remarkably accurate: experienced astronomers can evaluate a stellar magnitude within 0.1–0.2. The observer needs to compare the variable star with two nearby stars. Let V be the variable star, A a slightly brighter star, and B a slightly less bright star. The magnitude of V is estimated by interpolation after comparison with A and B. The observer notes his or her estimate in a scale from 1 to 5:

5: A is obviously much brighter than V.

4: A is noticeably brighter than V.

3: A is brighter than V, and the difference is clearly perceptible.

2: A and V at first seem of the same brightness, but a careful observation shows that A is in fact slightly brighter than V.

1: A and V seem to have the same luminosity, but A, from time to time, appears brighter.

Then, the same process is carried out comparing V with the less bright star B.

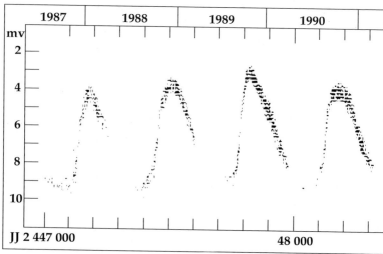

As an example, assume that the magnitude of A is mA = 8.2 and that of B is mB = 9.8. Our observation shows that comparing V to A corresponds to case No. 1, while comparing V to B corresponds to case No. 3. Generally speaking, we can write the magnitude of V as:

$$mV = \alpha\,\frac{(mB-mA)}{\beta + \alpha} + mA$$

with $\alpha = 1$ and $\beta = 3$ in our example, which gives:

$$mV = 1\,\frac{(9.8-8.2)}{3+1} + 8.2$$

$$mV = 8.6$$

Identifying a variable star and comparing it with comparison stars requires very detailed charts. As far as possible the observer will choose comparison stars of the same color as the variable star in order to make consistent measurements. Stars A and B must not be too far from each other and must already have a known and well-calibrated magnitude.

*In France, the quarterly bulletin (available to members) of the French Association of Variable Stars Observers (AFOEV) publishes annually over 65 000 stellar measurements and has a chart catalogue of nearly 800 stars. In the USA, contact the American Association of Variable Star Observers.*

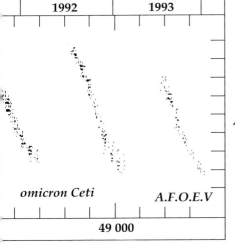

*The light curve of Mira Ceti between 1987 and 1993, established by members of the French Association of Variable Star Observers (AFOEV). The period is given in Julian days. From 1991 onwards, the curve shows gaps in observation due to the conjunction of the constellation Cetus with the Sun.* (© AFOEV.)

# VARIABLE STARS: THE R CrB GROUP

## History

R CrB is a very peculiar star, located in the Corona Borealis (The Northern Crown). Discovered by Pigott in 1795, it belongs to a group of thirty or so stars which undergo sudden magnitude variations, which are poorly understood. These stars are called R CrB type variable stars.

## Observation

These stars are usually observed at their maximum magnitude. R CrB has a magnitude of about 5.9–6 and can thus be seen with any instrument. But, in a sudden and unpredictable way, its luminosity drops to eighth magnitude within a few weeks (the ratio is 1600!). If this happened to the Sun, nothing would be left on Earth!

R CrB then falls to fourteenth magnitude, and takes a much longer time to return to its highest brightness, with less spectacular falls in between. It is said that during the 11 years between 1864 and 1875, R CrB

*Location charts for the star R CrB. On chart B, south is up, as in the field of the reflector (© AFOEV).*

1544+28 R Coronae Borealis (R CrB)
1900: 15 h 44 m 27 s   +28° 27.8'
1950: 15 h 46 m 30 s   +28° 18.5'
2000: 15 h 48 m 34 s   +28° 09.2'
RCB – mv 5.7 to 14.8 – irreg. – sp. C0,0 (F8pep)

annual precession
+2.47s -0.186'
eq. 1900.0

Chart from the *Photometric Atlas of Constellations* (A. Brun/AFOEV).

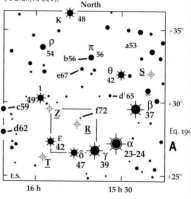

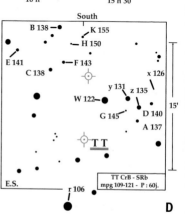

| R CrB | Period: none |
|---|---|
| Epoch 1950 coord: α: 15 h 44 m 27 s | **Spectra**: c F–G |
| δ: +28° 27.8′ | Const.: Corona Borealis (The Northern Crown) |
| Type: R CrB | |
| Mv: 5.8 to 14.8 | Favorable period: summer |

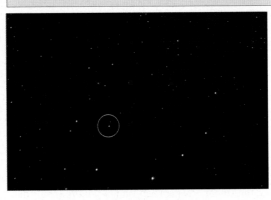

*R CrB (circled) at maximum brightness, near sixth magnitude* (© H. Burillier).

*Light curve for the star R CrB between 1983 and 1993. Amplitudes are chaotic, showing sudden drops in magnitude followed by long plateaux at maximum luminosity.* (© AFOEV.)

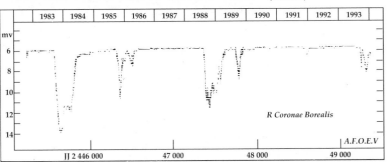

remained at maximum brightness ...What happens inside these stars?

R CrB stars are type G **supergiants**. Nowadays, the most probable explanation of their behavior is that these stars have an abnormally high carbon content, which, condensing in the atmosphere, could form sooty clouds diminishing the brightness of the star. These clouds would once again become gaseous as they fall towards the surface and disappear, and the star would then regain its normal brightness. Although this explanation satisfies most astronomers nowadays, it does not resolve all the mysteries surrounding these objects, in particular the mechanism by which carbon is ejected into the high atmosphere.

# SUPERNOVAE

Examples **THE CRAB NEBULA AND THE CYGNUS LOOP**

## History

In the universe, some massive stars (of eight solar masses or more) end their lives in a cataclysmic explosion, in which their envelopes are violently ejected and dispersed into the interstellar medium: these are supernovae. In the center, the stellar remnant is an extremely dense core, a rapidly rotating **neutron star**: a pulsar. The Crab Nebula (M1) is the remnant of a star which exploded in 1054 and was visible during the day for three weeks! Chinese annals relate the event: on the 4th of July, 1054, Min'Thuan-Lin wrote: "In the first year of Shi-ho period during the fifth Moon, the Tschi-Tschu day, a star appeared like a lighthouse near the star Tien-Kuan" [ζ Tau]. J. Bévis discovered its remnant in 1731.

The older Cygnus Loop is made up of gas, probably ejected 30 000 years ago, and is still expanding at a rate of 50 km/s. William Herschel discovered it in 1784.

## Observation

Using 7×50 binoculars the observer sees M1 as a blurred star, near ζ Tau. A particularly dark sky is needed to see the most luminous area of The Cygnus Loop (NGC 6960) using 11×80 binoculars. The amateur astronomer will need an aperture of at least 200 mm to make detailed observations of these two

*M1 ... or the cataclysmic death of a star*
(© Lick Observatory/*Ciel et Espace*).

| M1 | Apparent size: 5′×3′ |
|---|---|
| NGC 1952 | Absolute ⌀: 10 l.y. |
| Epoch 2000.0 coord: α: 05 h 34 m 5 s | Distance: 6500 l.y. |
| δ: +22° 01′ | Const.: Taurus |
| Mv: 9 | Favorable period: winter |

*The search for extra-galactic supernovae is a favorite with today's amateur astronomers. The supernova observed in M81 in April 1993 and discovered by a Spanish amateur astronomer is an example.*
(© A. Fujii/*Ciel et Espace*.)

| NGC 6960 |
|---|
| Epoch 2000.0 coord: α: 20 h 45 m 7 s |
| δ: +30° 43′ |
| Mv: 8 |
| Apparent size: 70′×6′ |
| Absolute ⌀: 100 l.y. |
| Distance: 1500 to 2500 l.y. |
| Const.: Cygnus |
| Favorable period: summer |

objects, and may do better to concentrate on NGC 6960, near the star 52 Cygni. This obviously bright area of the Loop can be made out using a 115/900 reflector with a UHC filter and a magnification of 45.

The appearance of a galactic supernova is a particularly rare phenomenon, and consistent observation of galaxies is a favorite activity of amateur hunters of supernovae. The discovery of a "new star" requires the making of location charts, the observation of magnitude fluctuations, plotting light curves, etc. *In France, the French Association of Variable Star Observers (AFOEV) coordinates these tasks, makes inventories and analyzes the meticulous work of its members.*

# Glossary

**Ablation**: a meteor entering the Earth's atmosphere is subject to intense heating, which results in erosion of its surface: this is ablation.

**Bok globules**: dark spherical clouds of dust and gas, associated with stellar formation zones.

**Cepheids**: a group of pulsating variable stars, whose prototype is the star δ Cephei.

**Conjunction**: the apparent location of two or more stellar bodies in the same region of the sky.

**Dwarf star**: a star of very small diameter (a few thousand kilometers) and very large mass (about that of the Sun) made up of degenerate matter (free-electron gas). Dwarf stars are the ultimate stage in the evolution of non-massive stars.

**Elongation**: the angular distance of a stellar body from the Sun, as measured from the Earth.

**Giant**: a luminous star of large radius and low density.

**Heliacal**: the rising or setting of a star simultaneously with the Sun.

**Local group**: the galaxy cluster to which our galaxy (called The Galaxy) belongs.

**Lyra (RR)**: old stars (variable) mainly found in globular clusters.

**Magellanic clouds**: two small galaxies in the austral sky, satellites of The Galaxy.

**Neutron star**: an extremely dense star (1 $cm^3$ of its matter can weigh 100 tonnes).

**Opposition**: said of two stars whose angular distance measured on the celestial sphere is 180°.

**Protostar**: interstellar matter in the phase of gravitational contraction, from which stars are born.

**Spectrum (spectral classification)**: the spectrum – monochromatic rays from the decomposition of light – differs according to a body's temperature. The Harvard Observatory Classification, adopted in the early twentieth century, classifies spectral types, from the hottest to the coldest, with the letters: O, B, A, F, G, K, M. Each type is then divided into 10, from 0 to 9. The Sun is of type G2.

**Sublimation**: passage from the solid to the gaseous state.

**Supergiants**: extremely luminous stars, of very low density and large diameter.

**Supernova**: a massive star which ends its evolution in a cataclysmic explosion.

**Tau (RV)**: supergiant variable stars characterised by their irregular luminosity fluctuations.

**Terminator**: the boundary between the illuminated and dark regions of the Moon or a planet.

**Variable star**: a star whose luminosity varies with time, either because of a companion (ecliptic variable), or because of its internal structure (intrinsic variable).

**White dwarf**: a small body of high temperature (some 10 000 K), with a low luminosity and high density (1 Earth's radius for 1 solar mass).

**Zodiac**: area of the celestial sphere spreading over 8° or so on each side of the ecliptic, and where the Moon, the Sun, and the planets (excepting Pluto) orbit.

# INDEX